Introduction to
Graph Neural Networks

Synthesis Lectures on Artificial Intelligence and Machine Learning

Editors

Ronald Brachman, *Jacobs Technion-Cornell Institute at Cornell Tech*
Francesca Rossi, *IBM Research AI*
Peter Stone, *University of Texas at Austin*

Introduction to Graph Neural Networks
Zhiyuan Liu and Jie Zhou
2020

Introduction to Logic Programming
Michael Genesereth and Vinay Chaudhri
2020

Federated Learning
Qiang Yang, Yang Liu, Yong Cheng, Yan Kang, and Tianjian Chen
2019

An Introduction to the Planning Domain Definition Language
Patrik Haslum, Nir Lipovetzky, Daniele Magazzeni, and Christina Muise
2019

Reasoning with Probabilistic and Deterministic Graphical Models: Exact Algorithms, Second Edition
Rina Dechter
2019

Learning and Decision-Making from Rank Data
Liron Xia
2019

Lifelong Machine Learning, Second Edition
Zhiyuan Chen and Bing Liu
2018

Adversarial Machine Learning
Yevgeniy Vorobeychik and Murat Kantarcioglu
2018

Strategic Voting
Reshef Meir
2018

Predicting Human Decision-Making: From Prediction to Action
Ariel Rosenfeld and Sarit Kraus
2018

Game Theory for Data Science: Eliciting Truthful Information
Boi Faltings and Goran Radanovic
2017

Multi-Objective Decision Making
Diederik M. Roijers and Shimon Whiteson
2017

Lifelong Machine Learning
Zhiyuan Chen and Bing Liu
2016

Statistical Relational Artificial Intelligence: Logic, Probability, and Computation
Luc De Raedt, Kristian Kersting, Sriraam Natarajan, and David Poole
2016

Representing and Reasoning with Qualitative Preferences: Tools and Applications
Ganesh Ram Santhanam, Samik Basu, and Vasant Honavar
2016

Metric Learning
Aurélien Bellet, Amaury Habrard, and Marc Sebban
2015

Graph-Based Semi-Supervised Learning
Amarnag Subramanya and Partha Pratim Talukdar
2014

Robot Learning from Human Teachers
Sonia Chernova and Andrea L. Thomaz
2014

General Game Playing
Michael Genesereth and Michael Thielscher
2014

Judgment Aggregation: A Primer
Davide Grossi and Gabriella Pigozzi
2014

An Introduction to Constraint-Based Temporal Reasoning
Roman Barták, Robert A. Morris, and K. Brent Venable
2014

Reasoning with Probabilistic and Deterministic Graphical Models: Exact Algorithms
Rina Dechter
2013

Introduction to Intelligent Systems in Traffic and Transportation
Ana L.C. Bazzan and Franziska Klügl
2013

A Concise Introduction to Models and Methods for Automated Planning
Hector Geffner and Blai Bonet
2013

Essential Principles for Autonomous Robotics
Henry Hexmoor
2013

Case-Based Reasoning: A Concise Introduction
Beatriz López
2013

Answer Set Solving in Practice
Martin Gebser, Roland Kaminski, Benjamin Kaufmann, and Torsten Schaub
2012

Planning with Markov Decision Processes: An AI Perspective
Mausam and Andrey Kolobov
2012

Active Learning
Burr Settles
2012

Computational Aspects of Cooperative Game Theory
Georgios Chalkiadakis, Edith Elkind, and Michael Wooldridge
2011

Representations and Techniques for 3D Object Recognition and Scene Interpretation
Derek Hoiem and Silvio Savarese
2011

A Short Introduction to Preferences: Between Artificial Intelligence and Social Choice
Francesca Rossi, Kristen Brent Venable, and Toby Walsh
2011

Human Computation
Edith Law and Luis von Ahn
2011

Trading Agents
Michael P. Wellman
2011

Visual Object Recognition
Kristen Grauman and Bastian Leibe
2011

Learning with Support Vector Machines
Colin Campbell and Yiming Ying
2011

Algorithms for Reinforcement Learning
Csaba Szepesvári
2010

Data Integration: The Relational Logic Approach
Michael Genesereth
2010

Markov Logic: An Interface Layer for Artificial Intelligence
Pedro Domingos and Daniel Lowd
2009

Introduction to Semi-Supervised Learning
Xiaojin Zhu and Andrew B.Goldberg
2009

Action Programming Languages
Michael Thielscher
2008

Representation Discovery using Harmonic Analysis
Sridhar Mahadevan
2008

Essentials of Game Theory: A Concise Multidisciplinary Introduction
Kevin Leyton-Brown and Yoav Shoham
2008

A Concise Introduction to Multiagent Systems and Distributed Artificial Intelligence
Nikos Vlassis
2007

Intelligent Autonomous Robotics: A Robot Soccer Case Study
Peter Stone
2007

Introduction to Graph Neural Networks

Zhiyuan Liu and Jie Zhou

ISBN: 978-3-031-00459-9   paperback
ISBN: 978-3-031-01587-8   ebook
ISBN: 978-3-031-00032-4   hardcover

DOI 10.1007/978-3-031-01587-8

A Publication in the Springer series
SYNTHESIS LECTURES ON ARTIFICIAL INTELLIGENCE AND MACHINE LEARNING

Lecture #45
Series Editors: Ronald Brachman, *Jacobs Technion–Cornell Institute at Cornell Tech*
                Francesca Rossi, *IBM Research AI*
                Peter Stone, *University of Texas at Austin*
Series ISSN
Synthesis Lectures on Artificial Intelligence and Machine Learning
Print 1939-4608   Electronic 1939-4616

Introduction to
Graph Neural Networks

Zhiyuan Liu and Jie Zhou
Tsinghua University

SYNTHESIS LECTURES ON ARTIFICIAL INTELLIGENCE AND MACHINE LEARNING #45

ABSTRACT

Graphs are useful data structures in complex real-life applications such as modeling physical systems, learning molecular fingerprints, controlling traffic networks, and recommending friends in social networks. However, these tasks require dealing with non-Euclidean graph data that contains rich relational information between elements and cannot be well handled by traditional deep learning models (e.g., convolutional neural networks (CNNs) or recurrent neural networks (RNNs)). Nodes in graphs usually contain useful feature information that cannot be well addressed in most unsupervised representation learning methods (e.g., network embedding methods). Graph neural networks (GNNs) are proposed to combine the feature information and the graph structure to learn better representations on graphs via feature propagation and aggregation. Due to its convincing performance and high interpretability, GNN has recently become a widely applied graph analysis tool.

This book provides a comprehensive introduction to the basic concepts, models, and applications of graph neural networks. It starts with the introduction of the vanilla GNN model. Then several variants of the vanilla model are introduced such as graph convolutional networks, graph recurrent networks, graph attention networks, graph residual networks, and several general frameworks. Variants for different graph types and advanced training methods are also included. As for the applications of GNNs, the book categorizes them into structural, non-structural, and other scenarios, and then it introduces several typical models on solving these tasks. Finally, the closing chapters provide GNN open resources and the outlook of several future directions.

KEYWORDS

deep graph learning, deep learning, graph neural network, graph analysis, graph convolutional network, graph recurrent network, graph residual network

Contents

Preface xv

Acknowledgments xvii

1 Introduction 1
   1.1   Motivations 1
       1.1.1   Convolutional Neural Networks 1
       1.1.2   Network Embedding 2
   1.2   Related Work 2

2 Basics of Math and Graph 5
   2.1   Linear Algebra 5
       2.1.1   Basic Concepts 5
       2.1.2   Eigendecomposition 7
       2.1.3   Singular Value Decomposition 8
   2.2   Probability Theory 8
       2.2.1   Basic Concepts and Formulas 8
       2.2.2   Probability Distributions 9
   2.3   Graph Theory 10
       2.3.1   Basic Concepts 10
       2.3.2   Algebra Representations of Graphs 10

3 Basics of Neural Networks 13
   3.1   Neuron 13
   3.2   Back Propagation 14
   3.3   Neural Networks 16

4 Vanilla Graph Neural Networks 19
   4.1   Introduction 19
   4.2   Model 19
   4.3   Limitations 21

5   Graph Convolutional Networks 23
   5.1   Spectral Methods 23
      5.1.1   Spectral Network 23
      5.1.2   ChebNet 24
      5.1.3   GCN 24
      5.1.4   AGCN 24
   5.2   Spatial Methods 25
      5.2.1   Neural FPs 25
      5.2.2   PATCHY-SAN 26
      5.2.3   DCNN 26
      5.2.4   DGCN 28
      5.2.5   LGCN 29
      5.2.6   MoNet 30
      5.2.7   GraphSAGE 31

6   Graph Recurrent Networks 33
   6.1   Gated Graph Neural Networks 33
   6.2   Tree LSTM 34
   6.3   Graph LSTM 35
   6.4   Sentence LSTM 36

7   Graph Attention Networks 39
   7.1   GAT 39
   7.2   GAAN 40

8   Graph Residual Networks 43
   8.1   Highway GCN 43
   8.2   Jump Knowledge Network 43
   8.3   DeepGCNs 45

9   Variants for Different Graph Types 47
   9.1   Directed Graphs 47
   9.2   Heterogeneous Graphs 48
   9.3   Graphs with Edge Information 49
   9.4   Dynamic Graphs 50
   9.5   Multi-Dimensional Graphs 51

10 Variants for Advanced Training Methods **53**
   10.1 Sampling 53
   10.2 Hierarchical Pooling 55
   10.3 Data Augmentation 55
   10.4 Unsupervised Training 56

11 General Frameworks **59**
   11.1 Message Passing Neural Networks 59
   11.2 Non-local Neural Networks 60
   11.3 Graph Networks 62

12 Applications – Structural Scenarios **67**
   12.1 Physics 67
   12.2 Chemistry and Biology 68
        12.2.1 Molecular Fingerprints 68
        12.2.2 Chemical Reaction Prediction 70
        12.2.3 Medication Recommendation 70
        12.2.4 Protein and Molecular Interaction Prediction ... 70
   12.3 Knowledge Graphs 71
        12.3.1 Knowledge Graph Completion 71
        12.3.2 Inductive Knowledge Graph Embedding 72
        12.3.3 Knowledge Graph Alignment 72
   12.4 Recommender Systems 73
        12.4.1 Matrix Completion 73
        12.4.2 Social Recommendation 74

13 Applications – Non-Structural Scenarios **75**
   13.1 Image 75
        13.1.1 Image Classification 75
        13.1.2 Visual Reasoning 77
        13.1.3 Semantic Segmentation 77
   13.2 Text 78
        13.2.1 Text Classification 78
        13.2.2 Sequence Labeling 79
        13.2.3 Neural Machine Translation 79
        13.2.4 Relation Extraction 79
        13.2.5 Event Extraction 81

13.2.6 Fact Verification 81
13.2.7 Other Applications 81

14 Applications – Other Scenarios **83**
14.1 Generative Models 83
14.2 Combinatorial Optimization 84

15 Open Resources **87**
15.1 Datasets 87
15.2 Implementations 88

16 Conclusion **91**

Bibliography **93**

Authors' Biographies **109**

Preface

Deep learning has achieved promising progress in many fields such as computer vision and natural language processing. The data in these tasks are usually represented in the Euclidean domain. However, many learning tasks require dealing with non-Euclidean graph data that contains rich relational information between elements, such as modeling physical systems, learning molecular fingerprints, predicting protein interface, etc. Graph neural networks (GNNs) are deep learning-based methods that operate on graph domains. Due to its convincing performance and high interpretability, GNN has recently been a widely applied graph analysis method.

The book provides a comprehensive introduction to the basic concepts, models, and applications of graph neural networks. It starts with the basics of mathematics and neural networks. In the first chapters, it gives an introduction to the basic concepts of GNNs, which aims to provide a general overview for readers. Then it introduces different variants of GNNs: graph convolutional networks, graph recurrent networks, graph attention networks, graph residual networks, and several general frameworks. These variants tend to generalize different deep learning techniques into graphs, such as convolutional neural network, recurrent neural network, attention mechanism, and skip connections. Further, the book introduces different applications of GNNs in structural scenarios (physics, chemistry, knowledge graph), non-structural scenarios (image, text) and other scenarios (generative models, combinatorial optimization). Finally, the book lists relevant datasets, open source platforms, and implementations of GNNs.

This book is organized as follows. After an overview in Chapter 1, we introduce some basic knowledge of math and graph theory in Chapter 2. We show the basics of neural networks in Chapter 3 and then give a brief introduction to the vanilla GNN in Chapter 4. Four types of models are introduced in Chapters 5, 6, 7, and 8, respectively. Other variants for different graph types and advanced training methods are introduced in Chapters 9 and 10. Then we propose several general GNN frameworks in Chapter 11. Applications of GNN in structural scenarios, nonstructural scenarios, and other scenarios are presented in Chapters 12, 13, and 14. And finally, we provide some open resources in Chapter 15 and conclude the book in Chapter 16.

Zhiyuan Liu and Jie Zhou
March 2020

Acknowledgments

We would like to thank those who contributed and gave advice in individual chapters:

Chapter 1: Ganqu Cui, Zhengyan Zhang

Chapter 2: Yushi Bai

Chapter 3: Yushi Bai

Chapter 4: Zhengyan Zhang

Chapter 9: Zhengyan Zhang, Ganqu Cui, Shengding Hu

Chapter 10: Ganqu Cui

Chapter 12: Ganqu Cui

Chapter 13: Ganqu Cui, Zhengyan Zhang

Chapter 14: Ganqu Cui, Zhengyan Zhang

Chapter 15: Yushi Bai, Shengding Hu

We would also thank those who provide feedback on the content of the book: Cheng Yang, Ruidong Wu, Chang Shu, Yufeng Du, and Jiayou Zhang.

Finally, we would like to thank all the editors, reviewers, and staff who helped with the publication of the book. Without you, this book would not have been possible.

Zhiyuan Liu and Jie Zhou
March 2020

CHAPTER 1

Introduction

Graphs are a kind of data structure which models a set of objects (nodes) and their relationships (edges). Recently, researches of analyzing graphs with machine learning have received more and more attention because of the great expressive power of graphs, i.e., graphs can be used as denotation of a large number of systems across various areas including social science (social networks) [Hamilton et al., 2017b, Kipf and Welling, 2017], natural science (physical systems [Battaglia et al., 2016, Sanchez et al., 2018] and protein-protein interaction networks [Fout et al., 2017]), knowledge graphs [Hamaguchi et al., 2017] and many other research areas [Khalil et al., 2017]. As a unique non-Euclidean data structure for machine learning, graph draws attention on analyses that focus on node classification, link prediction, and clustering. Graph neural networks (GNNs) are deep learning-based methods that operate on graph domain. Due to its convincing performance and high interpretability, GNN has been a widely applied graph analysis method recently. In the following paragraphs, we will illustrate the fundamental motivations of GNNs.

1.1 MOTIVATIONS

1.1.1 CONVOLUTIONAL NEURAL NETWORKS

Firstly, GNNs are motivated by convolutional neural networks (CNNs) LeCun et al. [1998]. CNNs is capable of extracting and composing multi-scale localized spatial features for features of high representation power, which have result in breakthroughs in almost all machine learning areas and the revolution of deep learning. As we go deeper into CNNs and graphs, we find the keys of CNNs: local connection, shared weights, and the use of multi-layer [LeCun et al., 2015]. These are also of great importance in solving problems of graph domain, because (1) graphs are the most typical locally connected structure, (2) shared weights reduce the computational cost compared with traditional spectral graph theory [Chung and Graham, 1997], (3) multi-layer structure is the key to deal with hierarchical patterns, which captures the features of various sizes. However, CNNs can only operate on regular Euclidean data like images (2D grid) and text (1D sequence), which can also be regarded as instances of graphs. Therefore, it is straightforward to think of finding the generalization of CNNs to graphs. As shown in Figure 1.1, it is hard to define localized convolutional filters and pooling operators, which hinders the transformation of CNN from Euclidean to non-Euclidean domain.

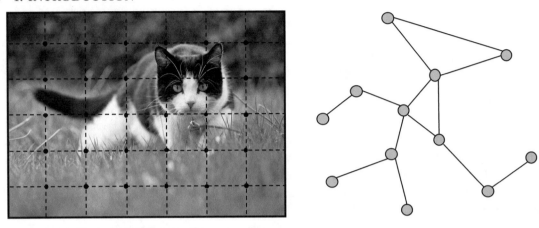

Figure 1.1: Left: image in Euclidean space. Right: graph in non-Euclidean space.

1.1.2   NETWORK EMBEDDING

The other motivation comes from *graph embedding* [Cai et al., 2018, Cui et al., 2018, Goyal and Ferrara, 2018, Hamilton et al., 2017a, Zhang et al., 2018a], which learns to represent graph nodes, edges, or subgraphs in low-dimensional vectors. In graph analysis, traditional machine learning approaches usually rely on hand-engineered features and are limited by its inflexibility and high cost. Following the idea of *representation learning* and the success of word embedding [Mikolov et al., 2013], DeepWalk [Perozzi et al., 2014], which is regarded as the first graph embedding method based on representation learning, applies SkipGram model [Mikolov et al., 2013] on the generated random walks. Similar approaches such as node2vec [Grover and Leskovec, 2016], LINE [Tang et al., 2015], and TADW [Yang et al., 2015b] also achieved breakthroughs. However, these methods suffer from two severe drawbacks [Hamilton et al., 2017a]. First, no parameters are shared between nodes in the encoder, which leads to computational inefficiency, since it means the number of parameters grows linearly with the number of nodes. Second, the direct embedding methods lack the ability of generalization, which means they cannot deal with dynamic graphs or be generalized to new graphs.

1.2   RELATED WORK

There exist several comprehensive reviews on GNNs. Monti et al. [2017] propose a unified framework, MoNet, to generalize CNN architectures to non-Euclidean domains (graphs and manifolds) and the framework could generalize several spectral methods on graphs [Atwood and Towsley, 2016, Kipf and Welling, 2017] as well as some models on manifolds [Boscaini et al., 2016, Masci et al., 2015]. Bronstein et al. [2017] provide a thorough review of geometric deep learning, which presents its problems, difficulties, solutions, applications, and future directions. Monti et al. [2017] and Bronstein et al. [2017] focus on generalizing convolutions to graphs or manifolds, while in this book we only focus on problems defined on graphs and we also investigate other mechanisms used in GNNs such as gate mechanism, attention mechanism, and skip connections. Gilmer et al. [2017] propose the message passing neural network (MPNN) which could generalize several GNN and graph convolutional network approaches. Wang et al. [2018b] propose the non-local neural network (NLNN) which unifies several "self-attention"-style methods. However, in the original paper, the model is not explicitly defined on graphs. Focusing on specific application domains, Gilmer et al. [2017] and Wang et al. [2018b] only give examples of how to generalize other models using their framework and they do not provide a review over other GNN models. Lee et al. [2018b] provides a review over graph attention models. Battaglia et al. [2018] propose the graph network (GN) framework which has a strong capability to generalize other models. However, the graph network model is highly abstract and Battaglia et al. [2018] only give a rough classification of the applications.

Zhang et al. [2018f] and Wu et al. [2019c] are two comprehensive survey papers on GNNs and they mainly focus on models of GNN. Wu et al. [2019c] categorize GNNs into four groups: recurrent graph neural networks (RecGNNs), convolutional graph neural networks (ConvGNNs), graph auto-encoders (GAEs), and spatial-temporal graph neural networks (STGNNs). Our book has a different taxonomy with Wu et al. [2019c]. We present graph recurrent networks in Chapter 6. Graph convolutional networks are introduced in Chapter 5 and a special variant of ConvGNNs, graph attention networks, are introduced in Chapter 7. We present the graph spatial-temporal networks in Section 9.4 as the models are usually used on dynamic graphs. We introduce graph auto-encoders in Section 10.4 as they are trained in an unsupervised fashion.

In this book, we provide a thorough introduction to different GNN models as well as a systematic taxonomy of the applications. To summarize, the major contents of this book are as follows.

- We provide a detailed review over existing GNN models. We introduce the original model, its variants and several general frameworks. We examine various models in this area and provide a unified representation to present different propagation steps in different models. One can easily make a distinction between different models using our representation by recognizing corresponding aggregators and updaters.

- We systematically categorize the applications and divide them into structural scenarios, non-structural scenarios, and other scenarios. For each scenario, we present several major applications and their corresponding methods.

CHAPTER 2

Basics of Math and Graph

2.1   LINEAR ALGEBRA

The language and concepts of linear algebra have come into widespread usage in many fields in computer science, and machine learning is no exception. A good comprehension of machine learning is based upon a thoroughly understanding of linear algebra. In this section, we will briefly review some important concepts and calculations in linear algebra, which are necessary for understanding the rest of the book. In this section, we review some basic concepts and calculations in linear algebra which are necessary for understanding the rest of the book.

2.1.1   BASIC CONCEPTS

- **Scalar**: A number.

- **Vector**: A column of ordered numbers, which can be expressed as the following:

$$\mathbf{x} = \begin{bmatrix} x_1 \\ x_2 \\ \vdots \\ x_n \end{bmatrix}. \tag{2.1}$$

The **norm** of a vector measures its length. The L_p norm is defined as follows:

$$||\mathbf{x}||_p = \left(\sum_{i=1}^{n} |x_i|^p \right)^{\frac{1}{p}}. \tag{2.2}$$

The L_1 norm, L_2 norm and L_∞ norm are often used in machine learning. The **L_1 norm** can be simplified as

$$||\mathbf{x}||_1 = \sum_{i=1}^{n} |x_i|. \tag{2.3}$$

In Euclidean space $\mathbb{R}^n$, the **L_2 norm** is used to measure the length of vectors, where

$$||\mathbf{x}||_2 = \sqrt{\sum_{i=1}^{n} x_i^2}. \tag{2.4}$$

The $\mathbf{L_\infty}$ **norm** is also called the max norm, as

$$||\mathbf{x}||_\infty = \max_i |x_i|.$$ (2.5)

With L_p norm, the **distance** of two vectors $\mathbf{x_1}$, $\mathbf{x_2}$ (where $\mathbf{x_1}$ and $\mathbf{x_2}$ are in the same linear space) can be defined as

$$\mathbf{D}_p(\mathbf{x_1}, \mathbf{x_2}) = ||\mathbf{x_1} - \mathbf{x_2}||_p.$$ (2.6)

A set of vectors $\mathbf{x_1}, \mathbf{x_2}, \cdots, \mathbf{x_m}$ are **linearly independent** if and only if there does not exist a set of scalars $\lambda_1, \lambda_2, \cdots, \lambda_m$, which are not all 0, such that

$$\lambda_1 \mathbf{x_1} + \lambda_2 \mathbf{x_2} + \cdots + \lambda_m \mathbf{x_m} = \mathbf{0}.$$ (2.7)

- **Matrix**: A two-dimensional array, which can be expressed as the following:

$$\mathbf{A} = \begin{bmatrix} a_{11} & a_{12} & \cdots & a_{1n} \\ a_{21} & a_{22} & \cdots & a_{2n} \\ \vdots & \vdots & \ddots & \vdots \\ a_{m1} & a_{m2} & \cdots & a_{mn} \end{bmatrix},$$ (2.8)

where $\mathbf{A} \in \mathbb{R}^{m \times n}$.

Given two matrices $\mathbf{A} \in \mathbb{R}^{m \times n}$ and $\mathbf{B} \in \mathbb{R}^{n \times p}$, the **matrix product** of $\mathbf{AB}$ can be denoted as $\mathbf{C} \in \mathbb{R}^{m \times p}$, where

$$\mathbf{C}_{ij} = \sum_{k=1}^{n} \mathbf{A}_{ik} \mathbf{B}_{kj}.$$ (2.9)

It can be proved that matrix product is associative but not necessarily commutative. In mathematical language,

$$(\mathbf{AB})\mathbf{C} = \mathbf{A}(\mathbf{BC})$$ (2.10)

holds for arbitrary matrices $\mathbf{A}, \mathbf{B}$, and $\mathbf{C}$ (presuming that the multiplication is legitimate). Yet

$$\mathbf{AB} = \mathbf{BA}$$ (2.11)

is not always true.

For each $n \times n$ square matrix $\mathbf{A}$, its **determinant** (also denoted as $|\mathbf{A}|$) is defined as

$$\det(\mathbf{A}) = \sum_{k_1 k_2 \cdots k_n} (-1)^{\tau(k_1 k_2 \cdots k_n)} a_{1k_1} a_{2k_2} \cdots a_{nk_n},$$ (2.12)

where $k_1 k_2 \cdots k_n$ is a permutation of $1, 2, \cdots, n$ and $\tau(k_1 k_2 \cdots k_n)$ is the **inversion number** of the permutation $k_1 k_2 \cdots k_n$, which is the number of **inverted sequence** in the permutation.

If matrix $\mathbf{A}$ is a square matrix, which means that $m = n$, the **inverse matrix** of $\mathbf{A}$ (denoted as $\mathbf{A}^{-1}$) satisfies that

$$\mathbf{A}^{-1}\mathbf{A} = \mathbf{I}, \tag{2.13}$$

where $\mathbf{I}$ is the $n \times n$ identity matrix. $\mathbf{A}^{-1}$ exists if and only if $|\mathbf{A}| \neq 0$.

The **transpose** of matrix $\mathbf{A}$ is represented as $\mathbf{A}^T$, where

$$\mathbf{A}^T_{ij} = \mathbf{A}_{ji}. \tag{2.14}$$

There is another frequently used product between matrices called **Hadamard product**. The Hadamard product of two matrices $\mathbf{A} \in \mathbb{R}^{m \times n}$ and $\mathbf{B} \in \mathbb{R}^{m \times n}$ is a matrix $\mathbf{C} \in \mathbb{R}^{m \times n}$, where

$$\mathbf{C}_{ij} = \mathbf{A}_{ij}\mathbf{B}_{ij}. \tag{2.15}$$

- **Tensor**: An array with arbitrary dimension. Most matrix operations can also be applied to tensors.

2.1.2   EIGENDECOMPOSITION

Let $\mathbf{A}$ be a matrix in $\mathbb{R}^{n \times n}$. A nonzero vector $\mathbf{v} \in \mathbb{C}^n$ is called an **eigenvector** of $\mathbf{A}$ if there exists such scalar $\lambda \in \mathbb{C}$ that

$$\mathbf{A}\mathbf{v} = \lambda\mathbf{v}. \tag{2.16}$$

Here scalar λ is an **eigenvalue** of $\mathbf{A}$ corresponding to the eigenvector $\mathbf{v}$. If matrix $\mathbf{A}$ has n eigenvectors $\{\mathbf{v}_1, \mathbf{v}_2, \cdots, \mathbf{v}_n\}$ that are linearly independent, corresponding to the eigenvalue $\{\lambda_1, \lambda_2, \cdots, \lambda_n\}$, then it can be deduced that

$$\mathbf{A}\begin{bmatrix} \mathbf{v}_1 & \mathbf{v}_2 & \cdots & \mathbf{v}_n \end{bmatrix} = \begin{bmatrix} \mathbf{v}_1 & \mathbf{v}_2 & \cdots & \mathbf{v}_n \end{bmatrix} \begin{bmatrix} \lambda_1 & & & \\ & \lambda_2 & & \\ & & \ddots & \\ & & & \lambda_n \end{bmatrix}. \tag{2.17}$$

Let $\mathbf{V} = \begin{bmatrix} \mathbf{v}_1 & \mathbf{v}_2 & \cdots & \mathbf{v}_n \end{bmatrix}$; then it is clear that $\mathbf{V}$ is an invertible matrix. We have the **eigendecomposition** of $\mathbf{A}$ (also called **diagonalization**)

$$\mathbf{A} = \mathbf{V}diag(\lambda)\mathbf{V}^{-1}. \tag{2.18}$$

It can also be written in the following form:

$$\mathbf{A} = \sum_{i=1}^{n} \lambda_i \mathbf{v}_i \mathbf{v}_i^T. \tag{2.19}$$

However, not all square matrices can be diagonalized in such form because a matrix may not have as many as n linear independent eigenvectors. Fortunately, it can be proved that every real symmetric matrix has an eigendecomposition.

2.1.3 SINGULAR VALUE DECOMPOSITION

As eigendecomposition can only be applied to certain matrices, we introduce the singular value decomposition, which is a generalization to all matrices.

First we need to introduce the concept of **singular value**. Let r denote the rank of $\mathbf{A}^T\mathbf{A}$, then there exist r positive scalars $\sigma_1 \geq \sigma_2 \geq \cdots \geq \sigma_r > 0$ such that for $1 \leq i \leq r$, $\mathbf{v}_i$ is an eigenvector of $\mathbf{A}^T\mathbf{A}$ with corresponding eigenvalue σ_i^2. Note that $\mathbf{v}_1, \mathbf{v}_2, \cdots, \mathbf{v}_r$ are linearly independent. The r positive scalars $\sigma_1, \sigma_2, \cdots, \sigma_r$ are called singular values of $\mathbf{A}$. Then we have the singular value decomposition

$$\mathbf{A} = U\Sigma V^T, \tag{2.20}$$

where $U \in \mathbb{R}^{m \times m}$ and V $(n \times n)$ are orthogonal matrices and Σ is an $m \times n$ matrix defined as follows:

$$\Sigma_{ij} = \begin{cases} \sigma_i & \text{if } i = j \leq r, \\ 0 & \text{otherwise.} \end{cases}$$

In fact, the column vectors of $\mathbf{U}$ are eigenvectors of $\mathbf{A}\mathbf{A}^T$, and the eigenvectors of $\mathbf{A}^T\mathbf{A}$ are made up of the the column vectors of $\mathbf{V}$.

2.2 PROBABILITY THEORY

Uncertainty is ubiquitous in the field of machine learning, thus we need to use probability theory to quantify and manipulate the uncertainty. In this section, we review some basic concepts and classic distributions in probability theory which are essential for understanding the rest of the book.

2.2.1 BASIC CONCEPTS AND FORMULAS

In probability theory, a **random variable** is a variable that has a random value. For instance, if we denote a random value by X, which has two possible values x_1 and x_2, then the probability of X equals to x_1 is $P(X = x_1)$. Clearly, the following equation remains true:

$$P(X = x_1) + P(X = x_2) = 1. \tag{2.21}$$

Suppose there is another random variable Y that has y_1 as a possible value. The probability that $X = x_1$ and $Y = y_1$ is written as $P(X = x_1, Y = y_1)$, which is called the **joint probability** of $X = x_1$ and $Y = y_1$.

Sometimes we need to know the relationship between random variables, like the probability of $X = x_1$ on the condition that $Y = y_1$, which can be written as $P(X = x_1|Y = y_1)$. We call this the **conditional probability** of $X = x_1$ given $Y = y_1$. With the concepts above, we can write the following two fundamental rules of probability theory:

$$P(X = x) = \sum_y P(X = x, Y = y), \tag{2.22}$$

$$P(X = x, Y = y) = P(Y = y|X = x)P(X = x). \tag{2.23}$$

The former is the **sum rule** while the latter is the **product rule**. Slightly modifying the form of product rule, we get another useful formula:

$$
\begin{aligned}
P(Y = y|X = x) &= \frac{P(X = x, Y = y)}{P(X = x)} \\
&= \frac{P(X = x|Y = y)P(Y = y)}{P(X = x)}
\end{aligned}
\tag{2.24}
$$

which is the famous **Bayes formula**. Note that it also holds for more than two variables:

$$P(X_i = x_i|Y = y) = \frac{P(Y = y|X_i = x_i)P(X_i = x_i)}{\sum_{j=1}^{n} P(Y = y|X_j = x_j)P(X_j = x_j)}. \tag{2.25}$$

Using product rule, we can deduce the **chain rule**:

$$
\begin{aligned}
&P(X_1 = x_1, \cdots, X_n = x_n) \\
&= P(X_1 = x_1) \prod_{i=2}^{n} P(X_i = x_i|X_1 = x_1, \cdots, X_{i-1} = x_{i-1}),
\end{aligned}
\tag{2.26}
$$

where $X_1, X_2, \cdots, X_n$ are n random variables.

The average value of some function $f(x)$ (where x is the value of a certain random variable) under a probability distribution $P(x)$ is called the **expectation** of $f(x)$. For a discrete distribution, it can be written as

$$\mathbb{E}[f(x)] = \sum_{x} P(x)f(x). \tag{2.27}$$

Usually, when $f(x) = x$, $\mathbb{E}[x]$ stands for the expectation of x.

To measure the dispersion level of $f(x)$ around its mean value $\mathbb{E}[f(x)]$, we introduce the **variance** of $f(x)$:

$$
\begin{aligned}
Var(f(x)) &= \mathbb{E}[(f(x) - \mathbb{E}[f(x)])^2] \\
&= \mathbb{E}[f(x)^2] - \mathbb{E}[f(x)]^2.
\end{aligned}
\tag{2.28}
$$

The **standard deviation** is the square root of variance. In some level, **covariance** expresses the degree to which two variables vary together:

$$Cov(f(x), g(y)) = \mathbb{E}[(f(x) - \mathbb{E}[f(x)])(g(y) - \mathbb{E}[g(y)])]. \tag{2.29}$$

Greater covariance indicates higher relevance between $f(x)$ and $g(y)$.

2.2.2   PROBABILITY DISTRIBUTIONS

Probability distributions describe the probability of a random variable or several random variables on every state. Several distributions that are useful in the area of machine learning are listed as follows.

- **Gaussian distribution**: it is also known as **normal distribution** and can be expressed as:

$$N\left(x|\mu,\sigma^2\right) = \sqrt{\frac{1}{2\pi\sigma^2}}\exp\left(-\frac{1}{2\sigma^2}(x-\mu)^2\right),\tag{2.30}$$

where μ is the mean of variable x and σ^2 is the variance.

- **Bernoulli distribution**: random variable X can either be 0 or 1, with a probability $P(X = 1) = p$. Then the distribution function is

$$P(X = x) = p^x(1-p)^{1-x}, x \in \{0, 1\}.\tag{2.31}$$

It is quite obvious that $E(X) = p$ and $Var(X) = p(1-p)$.

- **Binomial distribution**: repeat the Bernoulli experiment for N times and the times that X equals to 1 is denote by Y, then

$$P(Y = k) = \binom{N}{k}p^k(1-p)^{N-k}\tag{2.32}$$

is the Binomial distribution satisfying that $E(Y) = np$ and $Var(Y) = np(1-p)$.

- **Laplace distribution**: the Laplace distribution is described as

$$P(x|\mu,b) = \frac{1}{2b}\exp\left(-\frac{|x-\mu|}{b}\right).\tag{2.33}$$

2.3   GRAPH THEORY

Graphs are the basic subjects in the study of GNNs. Therefore, to get a comprehensive understanding of GNN, the fundamental graph theory is required.

2.3.1   BASIC CONCEPTS

A graph is often denoted by $G = (V, E)$, where V is the set of **vertices** and E is the set of **edges**. An edge $e = u, v$ has two **endpoints** u and v, which are said to be **joined** by e. In this case, u is called a **neighbor** of v, or in other words, these two vertices are **adjacent**. Note that an edge can either be **directed** or **undirected**. A graph is called a **directed graph** if all edges are directed or **undirected graph** if all edges are undirected. The **degree** of vertice v, denoted by $d(v)$, is the number of edges connected with v.

2.3.2   ALGEBRA REPRESENTATIONS OF GRAPHS

There are a few useful algebra representations for graphs, which are listed as follows.

- **Adjacency matrix**: for a simple graph $G = (V, E)$ with n-vertices, it can be described by an adjacency matrix $A \in \mathbb{R}^{n \times n}$, where

$$A_{ij} = \begin{cases} 1 & \text{if } \{v_i, v_j\} \in E \text{ and } i \neq j, \\ 0 & \text{otherwise.} \end{cases}$$

It is obvious that such matrix is a symmetric matrix when G is an undirected graph.

- **Degree matrix**: for a graph $G = (V, E)$ with n-vertices, its degree matrix $D \in \mathbb{R}^{n \times n}$ is a diagonal matrix, where

$$D_{ii} = d(v_i).$$

- **Laplacian matrix**: for a simple graph $G = (V, E)$ with n-vertices, if we consider all edges in G to be undirected, then its Laplacian matrix $L \in \mathbb{R}^{n \times n}$ can be defined as

$$L = D - A.$$

Thus, we have the elements:

$$L_{ij} = \begin{cases} d(v_i) & \text{if } i = j, \\ -1 & \text{if } \{v_i, v_j\} \in E \text{ and } i \neq j, \\ 0 & \text{otherwise.} \end{cases}$$

Note that the graph is considered to be an undirected graph for the adjacency matrix.

- **Symmetric normalized Laplacian**: the symmetric normalized Laplacian is defined as:

$$L^{sym} = D^{-\frac{1}{2}} L D^{-\frac{1}{2}}$$
$$= I - D^{-\frac{1}{2}} A D^{-\frac{1}{2}}.$$

The elements are given by:

$$L_{ij}^{sym} = \begin{cases} 1 & \text{if } i = j \text{ and } d(v_i) \neq 0, \\ -\dfrac{1}{\sqrt{d(v_i)d(v_j)}} & \text{if } \{v_i, v_j\} \in E \text{ and } i \neq j, \\ 0 & \text{otherwise.} \end{cases}$$

- **Random walk normalized Laplacian**: it is defined as:

$$L^{rw} = D^{-1} L = I - D^{-1} A.$$

The elements can be computed by:

$$L_{ij}^{rw} = \begin{cases} 1 & \text{if } i = j \text{ and } d(v_i) \neq 0, \\ -\dfrac{1}{d(v_i)} & \text{if } \{v_i, v_j\} \in E \text{ and } i \neq j, \\ 0 & \text{otherwise.} \end{cases}$$

- **Incidence matrix**: another common used matrix in representing a graph is incidence matrix. For a directed graph $G = (V, E)$ with n-vertices and m-edges, the corresponding incidence matrix is $M \in \mathbb{R}^{n \times m}$, where

$$
M_{ij} = \begin{cases} 1 & \text{if } \exists k \text{ s.t } e_j = \{v_i, v_k\}, \\ -1 & \text{if } \exists k \text{ s.t } e_j = \{v_k, v_i\}, \\ 0 & \text{otherwise.} \end{cases}
$$

For a undirected graph, the corresponding incidence matrix satisfies that

$$
M_{ij} = \begin{cases} 1 & \text{if } \exists k \text{ s.t } e_j = \{v_i, v_k\}, \\ 0 & \text{otherwise.} \end{cases}
$$

CHAPTER 3

Basics of Neural Networks

Neural networks are one of the most important models in machine learning. The structure of artificial neural networks, which consists of numerous neurons with connections to each other, bears great resemblance to that of biological neural networks. A neural network learns in the following way: initiated with random weights or values, the connections between neurons updates its weights or values by the back propagation algorithm repeatedly till the model performs rather precisely. In the end, the knowledge that a neural network learned is stored in the connections in a digital manner. Most of the researches on neural network try to change the way it learns (with different algorithms or different structures), aiming to improve the generalization ability of the model.

3.1 NEURON

The basic units of neural networks are **neurons,** which can receive a series of inputs and return the corresponding output. A classic neuron is as shown in Figure 3.1. Where the neuron receives n inputs $x_1, x_2, \cdots, x_n$ with corresponding weights $w_1, w_2, \cdots, w_n$ and an offset b. Then the weighted summation $y = \sum_{i=1}^{n} w_i x_i + b$ passes through an activation function f and the neuron returns the output $z = f(y)$. Note that the output will be the input of the next neuron. The **activation function** is a kind of function that maps a real number to a number between 0 and 1 (with rare exceptions), which represents the activation of the neuron, where 0 indicates deactivated and 1 indicates fully activated. Several useful activation functions are shown as follows.

- **Sigmoid Function** (Figure 3.2):

$$\sigma(x) = \frac{1}{1 + e^{-x}}. \tag{3.1}$$

- **Tanh Function** (Figure 3.3):

$$tanh(x) = \frac{e^x - e^{-x}}{e^x + e^{-x}}. \tag{3.2}$$

- **ReLU (Rectified Linear Unit)** (Figure 3.4):

$$ReLU(x) = \begin{cases} 0 & x \leq 0, \\ x & x > 0. \end{cases} \tag{3.3}$$

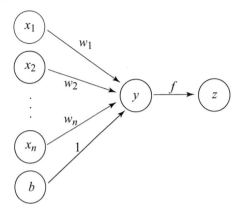

Figure 3.1: A classic neuron structure.

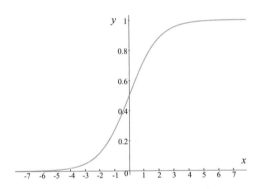

Figure 3.2: The Sigmoid function.

In fact, there are many other activation functions and each has its corresponding derivatives. But do remember that a good activation function is always smooth (which means that it is a continuous differentiable function) and easily calculated (in order to minimize the computational complexity of the neural network). During the training of a neural network, the choice of activation function is usually essential to the outcome.

3.2   BACK PROPAGATION

During the training of a neural network, the **back propagation algorithm** is most commonly used. It is an algorithm based on gradient descend to optimize the parameters in a model. Let's take the single neuron model illustrated above for an example. Suppose the optimization target for the output z is z_0, which will be approached by adjusting the parameters $w_1, w_2, \cdots, w_n, b$.

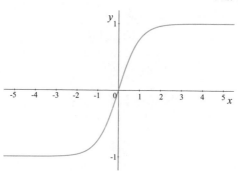

Figure 3.3: The Tanh function.

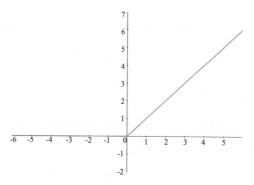

Figure 3.4: The ReLU (Rectified Linear Unit) function.

By the chain rule, we can deduce the derivative of z with respect to w_i and b:

$$\frac{\partial z}{\partial w_i} = \frac{\partial z}{\partial y}\frac{\partial y}{\partial w_i}$$
$$= \frac{\partial f(y)}{\partial y}x_i \qquad (3.4)$$

$$\frac{\partial z}{\partial b} = \frac{\partial z}{\partial y}\frac{\partial y}{\partial b}$$
$$= \frac{\partial f(y)}{\partial y}. \qquad (3.5)$$

With a learning rate of η, the update for each parameter will be:

$$\Delta w_i = \eta(z_0 - z)\frac{\partial z}{\partial w_i}$$
$$= \eta(z_0 - z)x_i\frac{\partial f(y)}{\partial y} \qquad (3.6)$$

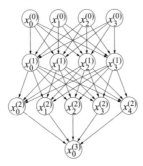

Figure 3.5: Feedforward neural network.

$$\Delta b = \eta(z_0 - z)\frac{\partial z}{\partial b}$$
$$= \eta(z_0 - z)\frac{\partial f(y)}{\partial y}. \tag{3.7}$$

In summary, the process of the back propagation consists of the following two steps.

- **Forward calculation**: given a set of parameters and an input, the neural network computes the values at each neuron in a forward order.

- **Backward propagation**: compute the error at each variable to be optimized, and update the parameters with their corresponding partial derivatives in a backward order.

The above two steps will go on repeatedly until the optimization target is acquired.

3.3   NEURAL NETWORKS

Recently, there is a booming development in the field of machine learning (especially deep learning), represented by the appearance of a variety of neural network structures. Though varying widely, the current neural network structures can be classified into several categories: feedforward neural networks, convolutional neural networks, recurrent neural networks, and GNNs.

- **Feedforward neural network**: The feedforward neural network (FNN) (Figure 3.5) is the first and simplest network architecture of artificial neural network. The FNN usually contains an input layer, several hidden layers, and an output layer. The feedforward neural network has a clear hierarchical structure, which always consists of multiple layers of neurons, and each layer is only connected to its neighbor layers. There are no loops in this network.

- **Convolutional neural network**: Convolutional neural networks (CNNs) are special versions of FNNs. FNNs are usually fully connected networks while CNNs preserve the local

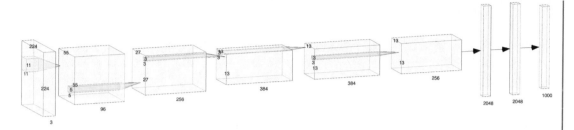

Figure 3.6: The AlexNet architecture from Krizhevsky et al. [2012].

connectivity. The CNN architecture usually contains convolutional layers, pooling layers, and several fully connected layers. There exist several classical CNN architectures such as LeNet5 [LeCun et al., 1998], AlexNet [Krizhevsky et al., 2012] (Figure 3.6), VGG [Simonyan and Zisserman, 2014], and GoogLeNet [Szegedy et al., 2015]. CNNs are widely used in the area of computer vision and proven to be effective in many other research fields.

- **Recurrent neural network**: In comparison with FNN, the neurons in recurrent neural network (RNN) receive not only signals and inputs from other neurons, but also its own historical information. The memory mechanism in recurrent neural network (RNN) help the model to process series data effectively. However, the RNN usually suffers from the problem of long-term dependencies [Bengio et al., 1994, Hochreiter et al., 2001]. Several variants are proposed to solve the problem by incorporating the gate mechanism such as GRU [Cho et al., 2014] and LSTM [Hochreiter and Schmidhuber, 1997]. The RNN is widely used in the area of speech and natural language processing.

- **Graph neural network**: The GNN is designed specifically to handle graph-structured data, such as social networks, molecular structures, knowledge graphs, etc. Detailed descriptions of GNNs will be covered in the later chapters of this book.

C H A P T E R   4

Vanilla Graph Neural Networks

In this section, we describe the vanilla GNNs proposed in Scarselli et al. [2009]. We also list the limitations of the vanilla GNN in representation capability and training efficiency. After this chapter we will talk about several variants of the vanilla GNN model.

4.1 INTRODUCTION

The concept of GNN was first proposed in Gori et al. [2005], Scarselli et al. [2004, 2009]. For simplicity, we will talk about the model proposed in Scarselli et al. [2009], which aims to extend existing neural networks for processing graph-structured data.

A node is naturally defined by its features and related nodes in the graph. The target of GNN is to learn a state embedding $\mathbf{h}_v \in \mathbb{R}^s$, which encodes the information of the neighborhood, for each node. The state embedding $\mathbf{h}_v$ is used to produce an output $\mathbf{o}_v$, such as the distribution of the predicted node label.

In Scarselli et al. [2009], a typical graph is illustrated in Figure 4.1. The vanilla GNN model deals with the undirected homogeneous graph where each node in the graph has its input features $\mathbf{x}_v$ and each edge may also have its features. The paper uses $co[v]$ and $ne[v]$ to denote the set of edges and neighbors of node v. For processing other more complicated graphs such as heterogeneous graphs, the corresponding variants of GNNs could be found in later chapters.

4.2 MODEL

Given the input features of nodes and edges, next we will talk about how the model obtains the node embedding $\mathbf{h}_v$ and the output embedding $\mathbf{o}_v$.

In order to update the node state according to the input neighborhood, there is a parametric function f, called *local transition function*, shared among all nodes. In order to produce the output of the node, there is a parametric function g, called *local output function*. Then, $\mathbf{h}_v$ and $\mathbf{o}_v$ are defined as follows:

$$\mathbf{h}_v = f(\mathbf{x}_v, \mathbf{x}_{co[v]}, \mathbf{h}_{ne[v]}, \mathbf{x}_{ne[v]}) \tag{4.1}$$

$$\mathbf{o}_v = g(\mathbf{h}_v, \mathbf{x}_v), \tag{4.2}$$

where $\mathbf{x}$ denotes the input feature and $\mathbf{h}$ denotes the hidden state. $co[v]$ is the set of edges connected to node v and $ne[v]$ is set of neighbors of node v. So that $\mathbf{x}_v, \mathbf{x}_{co[v]}, \mathbf{h}_{ne[v]}, \mathbf{x}_{ne[v]}$ are the features of v, the features of its edges, the states and the features of the nodes in the

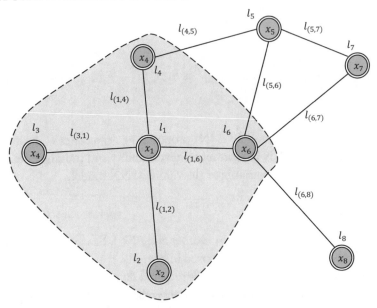

Figure 4.1: An example of the graph based on Scarselli et al. [2009].

neighborhood of v, respectively. In the example of node l_1 in Figure 4.1, $\mathbf{x}_{l_1}$ is the input feature of l_1. $co[l_1]$ contains edges $l_{(1,4)}, l_{(1,6)}, l_{(1,2)}$, and $l_{(3,1)}$. $ne[l_1]$ contains nodes l_2, l_3, l_4, and l_6.

Let $\mathbf{H}, \mathbf{O}, \mathbf{X}$, and $\mathbf{X}_N$ be the matrices constructed by stacking all the states, all the outputs, all the features, and all the node features, respectively. Then we have a compact form as:

$$\mathbf{H} = F(\mathbf{H}, \mathbf{X}) \tag{4.3}$$

$$\mathbf{O} = G(\mathbf{H}, \mathbf{X}_N) \tag{4.4}$$

where F, the *global transition function*, and G is the *global output function*. They are stacked versions of the local transition function f and the local output function g for all nodes in a graph, respectively. The value of $\mathbf{H}$ is the fixed point of Eq. (4.3) and is uniquely defined with the assumption that F is a contraction map.

With the suggestion of Banach's fixed point theorem [Khamsi and Kirk, 2011], GNN uses the following classic iterative scheme to compute the state:

$$\mathbf{H}^{t+1} = F(\mathbf{H}^t, \mathbf{X}), \tag{4.5}$$

where $\mathbf{H}^t$ denotes the tth iteration of $\mathbf{H}$. The dynamical system Eq. (4.5) converges exponentially fast to the solution of Eq. (4.3) for any initial value of $\mathbf{H}(0)$. Note that the computations described in f and g can be interpreted as the FNNs.

After the introduction of the framework of GNN, the next question is how to learn the parameters of the local transition function f and local output function g. With the target information ($\mathbf{t}_v$ for a specific node) for the supervision, the loss can be written as:

$$loss = \sum_{i=1}^{p} (\mathbf{t}_i - \mathbf{o}_i), \qquad (4.6)$$

where p is the number of supervised nodes. The learning algorithm is based on a gradient-descent strategy and is composed of the following steps.

- The states $\mathbf{h}_v^t$ are iteratively updated by Eq. (4.1) until a time step T. Then we obtain an approximate fixed point solution of Eq. (4.3): $\mathbf{H}(T) \approx \mathbf{H}$.

- The gradient of weights $\mathbf{W}$ is computed from the loss.

- The weights $\mathbf{W}$ are updated according to the gradient computed in the last step.

After running the algorithm, we can get a model trained for a specific supervised/semi-supervised task as well as hidden states of nodes in the graph. The vanilla GNN model provides an effective way to model graphic data and it is the first step toward incorporating neural networks into graph domain.

4.3   LIMITATIONS

Though experimental results showed that GNN is a powerful architecture for modeling structural data, there are still several limitations of the vanilla GNN.

- First, it is computationally inefficient to update the hidden states of nodes iteratively to get the fixed point. The model needs T steps of computation to approximate the fixed point. If relaxing the assumption of the fixed point, we can design a multi-layer GNN to get a stable representation of the node and its neighborhood.

- Second, vanilla GNN uses the same parameters in the iteration while most popular neural networks use different parameters in different layers, which serves as a hierarchical feature extraction method. Moreover, the update of node hidden states is a sequential process which can benefit from the RNN kernels like GRU and LSTM.

- Third, there are also some informative features on the edges which cannot be effectively modeled in the vanilla GNN. For example, the edges in the knowledge graph have the type of relations and the message propagation through different edges should be different according to their types. Besides, how to learn the hidden states of edges is also an important problem.

- Last, if T is pretty large, it is unsuitable to use the fixed points if we focus on the representation of nodes instead of graphs because the distribution of representation in the fixed point will be much more smooth in value and less informative for distinguishing each node.

Beyond the vanilla GNN, several variants are proposed to release these limitations. For example, Gated Graph Neural Network (GGNN) [Li et al., 2016] is proposed to solve the first problem. Relational GCN (R-GCN) [Schlichtkrull et al., 2018] is proposed to deal with directed graphs. More details could be found in the following chapters.

CHAPTER 5

Graph Convolutional Networks

In this chapter, we will talk about graph convolutional networks (GCNs), which aim to generalize convolutions to the graph domain. As convolutional neural networks (CNNs) have achieved great success in the area of deep learning, it is intuitive to define the convolution operation on graphs. Advances in this direction are often categorized as spectral approaches and spatial approaches. As there may have vast variants in each direction, we only list several classic models in this chapter.

5.1 SPECTRAL METHODS

Spectral approaches work with a spectral representation of the graphs. In this section we will talk about four classic models (Spectral Network, ChebNet, GCN, and AGCN).

5.1.1 SPECTRAL NETWORK

Bruna et al. [2014] propose the spectral network. The convolution operation is defined in the Fourier domain by computing the eigendecomposition of the graph Laplacian. The operation can be defined as the multiplication of a signal $\mathbf{x} \in \mathbb{R}^N$ (a scalar for each node) with a filter $\mathbf{g}_\theta = \text{diag}(\boldsymbol{\theta})$ parameterized by $\boldsymbol{\theta} \in \mathbb{R}^N$:

$$\mathbf{g}_\theta \star \mathbf{x} = \mathbf{U}\mathbf{g}_\theta(\Lambda)\mathbf{U}^T\mathbf{x}, \tag{5.1}$$

where $\mathbf{U}$ is the matrix of eigenvectors of the normalized graph Laplacian $\mathbf{L} = \mathbf{I}_N - \mathbf{D}^{-\frac{1}{2}}\mathbf{A}\mathbf{D}^{-\frac{1}{2}} = \mathbf{U}\Lambda\mathbf{U}^T$ ($\mathbf{D}$ is the degree matrix and $\mathbf{A}$ is the adjacency matrix of the graph), with a diagonal matrix of its eigenvalues Λ.

This operation results in potentially intense computations and non-spatially localized filters. Henaff et al. [2015] attempt to make the spectral filters spatially localized by introducing a parameterization with smooth coefficients.

5.1.2   CHEBNET

Hammond et al. [2011] suggest that $g_\theta(\Lambda)$ can be approximated by a truncated expansion in terms of Chebyshev polynomials $T_k(x)$ up to K^{th} order. Thus, the operation is:

$$g_\theta \star \mathbf{x} \approx \sum_{k=0}^{K} \theta_k T_k(\tilde{\mathbf{L}}) \mathbf{x} \tag{5.2}$$

with $\tilde{\mathbf{L}} = \frac{2}{\lambda_{max}} \mathbf{L} - \mathbf{I}_N$. λ_{max} denotes the largest eigenvalue of $\mathbf{L}$. $\theta \in \mathbb{R}^K$ is now a vector of Chebyshev coefficients. The Chebyshev polynomials are defined as $T_k(\mathbf{x}) = 2\mathbf{x}T_{k-1}(\mathbf{x}) - T_{k-2}(\mathbf{x})$, with $T_0(\mathbf{x}) = 1$ and $T_1(\mathbf{x}) = \mathbf{x}$. It can be observed that the operation is K-localized since it is a Kth-order polynomial in the Laplacian.

Defferrard et al. [2016] propose the ChebNet. It uses this K-localized convolution to define a convolutional neural network which could remove the need to compute the eigenvectors of the Laplacian.

5.1.3   GCN

Kipf and Welling [2017] limit the layer-wise convolution operation to $K = 1$ to alleviate the problem of overfitting on local neighborhood structures for graphs with very wide node degree distributions. It further approximates $\lambda_{max} \approx 2$ and the equation simplifies to:

$$g_{\theta'} \star \mathbf{x} \approx \theta_0' \mathbf{x} + \theta_1' (\mathbf{L} - \mathbf{I}_N) \mathbf{x} = \theta_0' \mathbf{x} - \theta_1' \mathbf{D}^{-\frac{1}{2}} \mathbf{A} \mathbf{D}^{-\frac{1}{2}} \mathbf{x} \tag{5.3}$$

with two free parameters θ_0' and θ_1'. After constraining the number of parameters with $\theta = \theta_0' = -\theta_1'$, we can obtain the following expression:

$$g_\theta \star \mathbf{x} \approx \theta \left(\mathbf{I}_N + \mathbf{D}^{-\frac{1}{2}} \mathbf{A} \mathbf{D}^{-\frac{1}{2}} \right) \mathbf{x}. \tag{5.4}$$

Note that stacking this operator could lead to numerical instabilities and exploding/vanishing gradients, Kipf and Welling [2017] introduce the *renormalization trick*: $\mathbf{I}_N + \mathbf{D}^{-\frac{1}{2}} \mathbf{A} \mathbf{D}^{-\frac{1}{2}} \to \tilde{\mathbf{D}}^{-\frac{1}{2}} \tilde{\mathbf{A}} \tilde{\mathbf{D}}^{-\frac{1}{2}}$, with $\tilde{\mathbf{A}} = \mathbf{A} + \mathbf{I}_N$ and $\tilde{\mathbf{D}}_{ii} = \sum_j \tilde{\mathbf{A}}_{ij}$. Finally, Kipf and Welling [2017] generalize the definition to a signal $\mathbf{X} \in \mathbb{R}^{N \times C}$ with C input channels and F filters for feature maps as follows:

$$\mathbf{Z} = \tilde{\mathbf{D}}^{-\frac{1}{2}} \tilde{\mathbf{A}} \tilde{\mathbf{D}}^{-\frac{1}{2}} \mathbf{X} \Theta \tag{5.5}$$

where $\Theta \in \mathbb{R}^{C \times F}$ is a matrix of filter parameters and $\mathbf{Z} \in \mathbb{R}^{N \times F}$ is the convolved signal matrix.

As a simplification of spectral methods, the GCN model could also be regarded as a spatial method as we listed in Section 5.2.

5.1.4   AGCN

All of the above models use the original graph structure to denote relations between nodes. However, there may have implicit relations between different nodes and the Adaptive Graph

Convolution Network (AGCN) is proposed to learn the underlying relations [Li et al., 2018b]. AGCN learns a "residual" graph Laplacian $\mathbf{L}_{res}$ and add it to the original Laplacian matrix:

$$\widehat{\mathbf{L}} = \mathbf{L} + \alpha \mathbf{L}_{res}, \tag{5.6}$$

where α is a parameter.

$\mathbf{L}_{res}$ is computed by a learned graph adjacency matrix $\widehat{\mathbf{A}}$

$$\mathbf{L}_{res} = \mathbf{I} - \widehat{\mathbf{D}}^{-\frac{1}{2}} \widehat{\mathbf{A}} \widehat{\mathbf{D}}^{-\frac{1}{2}}$$
$$\widehat{\mathbf{D}} = \text{degree}(\widehat{\mathbf{A}}), \tag{5.7}$$

and $\widehat{\mathbf{A}}$ is computed via a learned metric. The idea behind the adaptive metric is that Euclidean distance is not suitable for graph structured data and the metric should be adaptive to the task and input features. AGCN uses the generalized Mahalanobis distance:

$$D(\mathbf{x}_i, \mathbf{x}_j) = \sqrt{(\mathbf{x}_i - \mathbf{x}_j)^T \mathbf{M}(\mathbf{x}_i - \mathbf{x}_j)}, \tag{5.8}$$

where $\mathbf{M}$ is a learned parameter that satisfies $\mathbf{M} = \mathbf{W}_d \mathbf{W}_d^T$. $\mathbf{W}_d$ is the transform basis to the adaptive space. Then AGCN calucates the Gaussian kernel and normalize G to obtain the dense adjacency matrix $\widehat{\mathbf{A}}$:

$$G_{x_i, x_j} = \exp\left(-D(\mathbf{x}_i, \mathbf{x}_j) / (2\sigma^2)\right). \tag{5.9}$$

5.2   SPATIAL METHODS

In all of the spectral approaches mentioned above, the learned filters depend on the Laplacian eigenbasis, which depends on the graph structure. That means a model trained on a specific structure could not be directly applied to a graph with a different structure.

In contrast, spatial approaches define convolutions directly on the graph, operating on spatially close neighbors. The major challenge of spatial approaches is defining the convolution operation with differently sized neighborhoods and maintaining the local invariance of CNNs.

5.2.1   NEURAL FPS

Duvenaud et al. [2015] use different weight matrices for nodes with different degrees,

$$\mathbf{x} = \mathbf{h}_v^{t-1} + \sum_{i=1}^{|N_v|} \mathbf{h}_i^{t-1}$$
$$\mathbf{h}_v^t = \sigma\left(\mathbf{x} \mathbf{W}_t^{|N_v|}\right), \tag{5.10}$$

where $\mathbf{W}_t^{|N_v|}$ is the weight matrix for nodes with degree $|N_v|$ at layer t, N_v denotes the set of neighbors of node v, $\mathbf{h}_v^t$ is the embedding of node v at layer t. We can see from the equations that the model first adds the embeddings from itself as well as its neighbors, then it uses $\mathbf{W}_t^{|N_v|}$ to do the transformation. The model defines different matrices $\mathbf{W}_t^{|N_v|}$ for nodes with different degrees. And the main drawback of the method is that it cannot be applied to large-scale graphs with more node degrees.

5.2.2   PATCHY-SAN

The PATCHY-SAN model [Niepert et al., 2016] first selects and normalizes exactly k neighbors for each node. Then the normalized neighborhood serves as the receptive filed and the convolutional operation is applied. In detail, the method has four steps.

Node Sequence Selection. This method does not process with all nodes in the graph. Instead it selects a sequence of nodes for processing. It first uses a graph labeling procedure to get the order of the nodes and obtain the sequence of the nodes. Then the method uses a stride s to select nodes from the sequence until a number of w nodes are selected.

Neighborhood Assembly. In this step, the receptive fields of nodes selected from last step are constructed. The neighbors of each node are the candidates and the model uses a simple breadth-first search to collect k neighbors for each node. It first extracts the 1-hop neighbors of the node, then it considers high-order neighbors until the total number of k neighbors are extracted.

Graph Normalization. In this step, the algorithm aims to give an order to nodes in the receptive field, so that this step maps from the unordered graph space to a vector space. This is the most important step and the idea behind this step is to assign nodes from two different graphs similar relative positions if they have similar structural roles. More details of this algorithm could be found in Niepert et al. [2016].

Convolutional Architecture. After the receptive fields are normalized in last step, CNN architectures can be used. The normalized neighborhoods serve as receptive fields and node and edge attributes are regarded as channels.

An illustration of this model could be found in Figure 5.1. This method tries to convert the graph learning problem to the traditional euclidean data learning problem.

5.2.3   DCNN

Atwood and Towsley [2016] propose the diffusion-convolutional neural networks (DCNNs). Transition matrices are used to define the neighborhood for nodes in DCNN. For node classification, it has

$$\mathbf{H} = \sigma \left(\mathbf{W}^c \odot \mathbf{P}^* \mathbf{X} \right), \tag{5.11}$$

where $\mathbf{X}$ is an $N \times F$ tensor of input features (N is the number of nodes and F is the number of features). $\mathbf{P}^*$ is an $N \times K \times N$ tensor which contains the power series $\{\mathbf{P}, \mathbf{P}^2, ..., \mathbf{P}^K\}$ of matrix $\mathbf{P}$. And $\mathbf{P}$ is the degree-normalized transition matrix from the graphs adjacency matrix $\mathbf{A}$. Each entity is transformed to a diffusion convolutional representation which is a $K \times F$

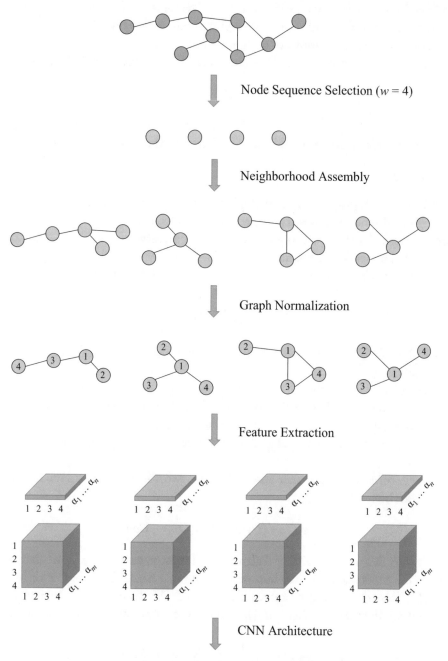

Figure 5.1: An illustration of the architecture of PATCHY-SAN. Note the figure is only for illustration and it is not the true algorithm output.

matrix defined by K hops of graph diffusion over F features. And then it will be defined by a $K \times F$ weight matrix and a nonlinear activation function σ. Finally, $\mathbf{H}$ (which is $N \times K \times F$) denotes the diffusion representations of each node in the graph.

As for graph classification, DCNN simply takes the average of nodes' representation,

$$\mathbf{H} = \sigma \left(\mathbf{W}^c \odot 1_N^T \mathbf{P}^* \mathbf{X}/N \right) \tag{5.12}$$

and 1_N here is an $N \times 1$ vector of ones. DCNN can also be applied to edge classification tasks, which requires converting edges to nodes and augmenting the adjacency matrix.

5.2.4 DGCN

Zhuang and Ma [2018] propose the dual graph convolutional network (DGCN) to jointly consider the local consistency and global consistency on graphs. It uses two convolutional networks to capture the local/global consistency and adopts an unsupervised loss to ensemble them. The first convolutional network is the same as Eq. (5.5). And the second network replaces the adjacency matrix with positive pointwise mutual information (PPMI) matrix:

$$\mathbf{H}' = \sigma \left(\mathbf{D}_P^{-\frac{1}{2}} \mathbf{X}_P \mathbf{D}_P^{-\frac{1}{2}} \mathbf{H}\Theta \right), \tag{5.13}$$

where $\mathbf{X}_P$ is the PPMI matrix and $\mathbf{D}_P$ is the diagonal degree matrix of $\mathbf{X}_P$, σ is a nonlinear activation function.

The motivations of jointly using the two perspectives are: (1) Eq. (5.5) models the local consistency, which indicates that nearby nodes may have similar labels, and (2) Eq. (5.13) models the global consistency which assumes that nodes with similar context may have similar labels. The local consistency convolution and global consistency convolution are named as $Conv_A$ and $Conv_P$.

Zhuang and Ma [2018] further ensemble these two convolutions via the final loss function. It can be written as:

$$L = L_0(Conv_A) + \lambda(t)L_{reg}(Conv_A, Conv_P). \tag{5.14}$$

$\lambda(t)$ is the dynamic weight to balance the importance of these two loss functions. $L_0(Conv_A)$ is the supervised loss function with given node labels. If we have c different labels to predict, Z^A denotes the output matrix of $Conv_A$ and $\widehat{Z}^A$ denotes the output of Z^A after a softmax operation, then the loss $L_0(Conv_A)$, which is the cross-entropy error, can be written as:

$$L_0(Conv_A) = -\frac{1}{|y_L|} \sum_{l \in y_L} \sum_{i=1}^{c} Y_{l,i} \ln \left(\widehat{Z}_{l,i}^A \right), \tag{5.15}$$

where y_L is the set of training data indices and Y is the ground truth.

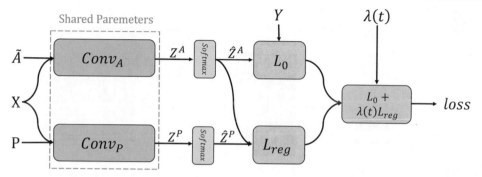

Figure 5.2: Architecture of the dual graph convolutional network (DGCN) model.

The calculation of L_{reg} can be written as:

$$L_{reg}(Conv_A, Conv_P) = \frac{1}{n} \sum_{i=1}^{n} \left\| \widehat{Z}_{i,:}^{P} - \widehat{Z}_{i,:}^{A} \right\|^2, \quad (5.16)$$

where $\widehat{Z}^P$ denotes the output of $Conv_P$ after the softmax operation. Thus, $L_{reg}(Conv_A, Conv_P)$ is the unsupervised loss function to measure the differences between $\widehat{Z}^P$ and $\widehat{Z}^A$. As a result, the architecture of this model is shown in Figure 5.2.

5.2.5   LGCN

Gao et al. [2018] propose the learnable graph convolutional networks (LGCN). The network is based on the learnable graph convolutional layer (LGCL) and the sub-graph training strategy. We will give the details of the learnable graph convolutional layer in this section.

LGCL leverages CNNs as aggregators. It performs max pooling on nodes' neighborhood matrices to get top-k feature elements and then applies 1-D CNN to compute hidden representations. The propagation step of LGCL is formulated as:

$$\begin{aligned}
\widehat{H}_t &= g\left(H_t, A, k\right) \\
H_{t+1} &= c\left(\widehat{H}_t\right),
\end{aligned} \quad (5.17)$$

where A is the adjacency matrix, $g(\cdot)$ is the k-largest node selection operation, and $c(\cdot)$ denotes the regular 1-D CNN.

The model uses the k-largest node selection operation to gather information for each node. For a given node x, the features of its neighbors are firstly gathered; suppose it has n neighbors and each node has c features, then a matrix $M \in \mathbb{R}^{n \times c}$ is obtained. If n is less than k, then M is padded with columns of zeros. Then the k-largest node selection is conducted that we rank the values in each column and select the top-k values. After that, the embedding of the node x is inserted into the first row of the matrix and finally we get a matrix $\widehat{M} \in \mathbb{R}^{(k+1) \times c}$.

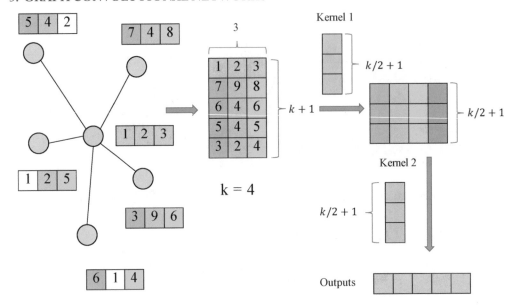

Figure 5.3: An example of the learnable graph convolutional layer (LGCL). Each node has three three features and this layer selects $k = 4$ neighbors. The node has five neighbors and four of them are selected. The k-largest node selection procedure is shown in the left part and four largest values are selected in each column. Then a 1-D CNN is performed to get the final output.

After the matrix $\widehat{M}$ is obtained, then the model uses the regular 1-D CNN to aggregate the features. The function $c(\cdot)$ should take a matrix of $N \times (k + 1) \times C$ as input and output a matrix of dimension $N \times D$ or $N \times 1 \times D$. Figure 5.3 gives an example of the LGCL.

5.2.6   MONET

Monti et al. [2017] propose a spatial-domain model (MoNet) on non-Euclidean domains which could generalize several previous techniques. The Geodesic CNN (GCNN) [Masci et al., 2015] and Anisotropic CNN (ACNN) [Boscaini et al., 2016] on manifolds or GCN [Kipf and Welling, 2017] and DCNN [Atwood and Towsley, 2016] on graphs could be formulated as particular instances of MoNet.

We use x to denote the node in the graph and $y \in N_x$ to denote the neighbor node of x. The MoNet model computes the *pseudo-coordinates* $\mathbf{u}(x, y)$ between the node and its neighbor and uses a weighting function among these coordinates:

$$D_j(x)f = \sum_{y \in N_x} w_j(\mathbf{u}(x, y)) f(y), \qquad (5.18)$$

Table 5.1: Different settings for different methods in the MoNet framework

Method	$\mathbf{u}(x, y)$	Weight Function $w_j(\mathbf{u})$
CNN	$\mathbf{x}(x, y) = \mathbf{x}(y) - \mathbf{x}(x)$	$\delta(\mathbf{u} - \bar{\mathbf{u}}_j)$
GCN	$\deg(x), \deg(y)$	$(1 - \|1 - \frac{1}{\sqrt{u_1}}\|) (1 - \frac{1}{\sqrt{u_2}}\|)$
DCNN	$p^0(x, y), \ldots p^{r-1}(x, y)$	$id(u_j)$

where the parameters are $\mathbf{w}_{\Theta}(\mathbf{u}) = (w_1(\mathbf{u}), \ldots, w_J(\mathbf{u}))$ and J represents the size of the extracted patch. Then a spatial generalization of the convolution on non-Euclidean domains is defined as:

$$(f \star g)(x) = \sum_{j=1}^{J} g_j D_j(x) f. \tag{5.19}$$

Then other methods can be regarded as a special case with different coordinates $\mathbf{u}$ and weight functions $\mathbf{w}(\mathbf{u})$. As we only focus on deep learning on graphs, we list several settings in Table 5.1. More details could be found in the original paper.

5.2.7 GRAPHSAGE

Hamilton et al. [2017b] propose the GraphSAGE, a general inductive framework. The framework generates embeddings by sampling and aggregating features from a node's local neighborhood. The propagation step of GraphSAGE is:

$$\begin{aligned} \mathbf{h}_{N_v}^t &= \text{AGGREGATE}_t \left(\{\mathbf{h}_u^{t-1}, \forall u \in N_v\} \right) \\ \mathbf{h}_v^t &= \sigma \left(\mathbf{W}^t \cdot \left[\mathbf{h}_v^{t-1} \| \mathbf{h}_{N_v}^t \right] \right), \end{aligned} \tag{5.20}$$

where $\mathbf{W}^t$ is the parameter at layer t.

However, Hamilton et al. [2017b] do not utilize the full set of neighbors in Eq. (5.20) but a fixed-size set of neighbors by uniformly sampling. The AGGREGATE function can have various forms. And Hamilton et al. [2017b] suggest three aggregator functions.

- **Mean aggregator.** It could be viewed as an approximation of the convolutional operation from the transductive GCN framework [Kipf and Welling, 2017], so that the inductive version of the GCN variant could be derived by

$$\mathbf{h}_v^t = \sigma \left(\mathbf{W} \cdot \text{MEAN}(\{\mathbf{h}_v^{t-1}\} \cup \{\mathbf{h}_u^{t-1}, \forall u \in N_v\}) \right). \tag{5.21}$$

The mean aggregator is different from other aggregators because it does not perform the concatenation operation which concatenates $\mathbf{h}_v^{t-1}$ and $\mathbf{h}_{N_v}^t$ in Eq. (5.20). It could be viewed as a form of "skip connection" [He et al., 2016b] and could achieve better performance.

- **LSTM aggregator.** Hamilton et al. [2017b] also use an LSTM-based aggregator which has a larger expressive capability. However, LSTMs process inputs in a sequential manner so that they are not permutation invariant. Hamilton et al. [2017b] adapt LSTMs to operate on an unordered set by permutating node's neighbors.

- **Pooling aggregator.** In the pooling aggregator, each neighbor's hidden state is fed through a fully connected layer and then a max-pooling operation is applied to the set of the node's neighbors:

$$\mathbf{h}^t_{N_v} = \max\left(\left\{\sigma\left(\mathbf{W}_{\text{pool}}\mathbf{h}^{t-1}_u + \mathbf{b}\right), \forall u \in N_v\right\}\right). \tag{5.22}$$

Note that any symmetric functions could be used in place of the max-pooling operation here.

To learn better representations, GraphSAGE further proposes an unsupervised loss function which encourages nearby nodes to have similar representations while distant nodes have different representations:

$$J_G(\mathbf{z}_u) = -\log\left(\sigma\left(\mathbf{z}^T_u \mathbf{z}_v\right)\right) - Q \cdot E_{v_n \sim P_n(v)}\log\left(\sigma\left(-\mathbf{z}^T_u \mathbf{z}_{v_n}\right)\right), \tag{5.23}$$

where v is the neighbor of node u and P_n is a negative sampling distribution. Q is the number of negative samples.

CHAPTER 6

Graph Recurrent Networks

There is another trend to use the gate mechanism from RNNs like GRU [Cho et al., 2014] or LSTM [Hochreiter and Schmidhuber, 1997] in the propagation step to diminish the restrictions from the vanilla GNN model and improve the effectiveness of the long-term information propagation across the graph. In this chapter, we will talk about several variants and we call them Graph Recurrent Networks (GRNs).

6.1 GATED GRAPH NEURAL NETWORKS

Li et al. [2016] propose the GGNN which uses the Gate Recurrent Units (GRU) in the propagation step. It unrolls the recurrent neural network for a fixed number of T steps and backpropagates through time to compute gradients.

Specifically, the basic recurrence of the propagation model is

$$
\begin{aligned}
\mathbf{a}_v^t &= \mathbf{A}_v^T \left[\mathbf{h}_1^{t-1} \ldots \mathbf{h}_N^{t-1} \right]^T + \mathbf{b} \\
\mathbf{z}_v^t &= \sigma \left(\mathbf{W}^z \mathbf{a}_v^t + \mathbf{U}^z \mathbf{h}_v^{t-1} \right) \\
\mathbf{r}_v^t &= \sigma \left(\mathbf{W}^r \mathbf{a}_v^t + \mathbf{U}^r \mathbf{h}_v^{t-1} \right) \\
\widetilde{\mathbf{h}}_v^t &= \tanh \left(\mathbf{W} \mathbf{a}_v^t + \mathbf{U} \left(\mathbf{r}_v^t \odot \mathbf{h}_v^{t-1} \right) \right) \\
\mathbf{h}_v^t &= \left(1 - \mathbf{z}_v^t \right) \odot \mathbf{h}_v^{t-1} + \mathbf{z}_v^t \odot \widetilde{\mathbf{h}}_v^t.
\end{aligned}
\tag{6.1}
$$

The node v first aggregates message from its neighbors, where $\mathbf{A}_v$ is the sub-matrix of the graph adjacency matrix $\mathbf{A}$ and denotes the connection of node v with its neighbors. The GRU-like update functions use information from each node's neighbors and from the previous timestep to update node's hidden state. Vector $\mathbf{a}$ gathers the neighborhood information of node v, $\mathbf{z}$, and $\mathbf{r}$ are the update and reset gates, $\odot$ is the Hardamard product operation.

The GGNN model is designed for problems defined on graphs which require outputting sequences while existing models focus on producing single outputs such as node-level or graph-level classifications.

Li et al. [2016] further propose Gated Graph Sequence Neural Networks (GGS-NNs) which uses several GGNNs to produce an output sequence $\mathbf{o}^{(1)} \ldots \mathbf{o}^{(K)}$. As shown in Figure 6.1, for the kth output step, the matrix of node annotations is denoted as $\mathbf{X}^{(k)}$. Two GGNNs are used in this architecture: (1) $F_o^{(k)}$ for predicting $\mathbf{o}^{(k)}$ from $\mathbf{X}^{(k)}$ and (2) $F_x^{(k)}$ for predicting $\mathbf{X}^{(k+1)}$ from $\mathbf{X}^{(k)}$. We use $\mathbf{H}^{(k,t)}$ to denote the t-th propagation step of the k-th output step. The value

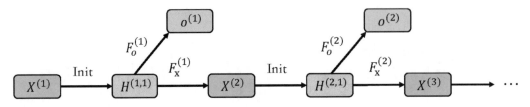

Figure 6.1: Architecture of Gated Graph Sequence Neural Network models.

of $\mathbf{H}^{(k,1)}$ at each step k is initialized by $\mathbf{X}^{(k)}$. The value of $\mathbf{H}^{(t,1)}$ at each step t is initialized by $\mathbf{X}^{(t)}$. $F_o^{(k)}$ and $F_x^{(k)}$ can be different models or share the same parameters.

The model is used on the bAbI task as well as the program verification task and has demonstrated its effectiveness.

6.2   TREE LSTM

LSTMs are also used in a similar way as GRU through the propagation process based on a tree or a graph.

Tai et al. [2015] propose two extensions to the basic LSTM architecture: the *Child-Sum Tree-LSTM* and the *N-ary Tree-LSTM*. Like in standard LSTM units, each Tree-LSTM unit (indexed by v) contains input and output gates $\mathbf{i}_v$ and $\mathbf{o}_v$, a memory cell $\mathbf{c}_v$, and hidden state $\mathbf{h}_v$. The Tree-LSTM unit abandons the single forget gate but uses a forget gate $\mathbf{f}_{vk}$ for each child k, allowing node v to aggregate information from its child accordingly. The equations for the Child-Sum Tree-LSTM are:

$$\widetilde{\mathbf{h}}_v^{t-1} = \sum_{k \in N_v} \mathbf{h}_k^{t-1}$$

$$\mathbf{i}_v^t = \sigma\left(\mathbf{W}^i \mathbf{x}_v^t + \mathbf{U}^i \widetilde{\mathbf{h}}_v^{t-1} + \mathbf{b}^i\right)$$

$$\mathbf{f}_{vk}^t = \sigma\left(\mathbf{W}^f \mathbf{x}_v^t + \mathbf{U}^f \mathbf{h}_k^{t-1} + \mathbf{b}^f\right)$$

$$\mathbf{o}_v^t = \sigma\left(\mathbf{W}^o \mathbf{x}_v^t + \mathbf{U}^o \widetilde{\mathbf{h}}_v^{t-1} + \mathbf{b}^o\right) \tag{6.2}$$

$$\mathbf{u}_v^t = \tanh\left(\mathbf{W}^u \mathbf{x}_v^t + \mathbf{U}^u \widetilde{\mathbf{h}}_v^{t-1} + \mathbf{b}^u\right)$$

$$\mathbf{c}_v^t = \mathbf{i}_v^t \odot \mathbf{u}_v^t + \sum_{k \in N_v} \mathbf{f}_{vk}^t \odot \mathbf{c}_k^{t-1}$$

$$\mathbf{h}_v^t = \mathbf{o}_v^t \odot \tanh(\mathbf{c}_v^t),$$

where $\mathbf{x}_v^t$ is the input vector at time t in the standard LSTM setting, $\odot$ is the Hardamard product operation.

If the number of children of each node in a tree is at most K and the children can be ordered from 1 to K, then the N-array Tree-LSTM can be applied. For node v, $\mathbf{h}_{vk}^t$ and $\mathbf{c}_{vk}^t$ denote the hidden state and memory cell of its k-th child at time t, respectively. The transition equations are the following:

$$\mathbf{i}_v^t = \sigma\left(\mathbf{W}^i \mathbf{x}_v^t + \sum_{l=1}^{K} \mathbf{U}_l^i \mathbf{h}_{vl}^{t-1} + \mathbf{b}^i\right)$$

$$\mathbf{f}_{vk}^t = \sigma\left(\mathbf{W}^f \mathbf{x}_v^t + \sum_{l=1}^{K} \mathbf{U}_{kl}^f \mathbf{h}_{vl}^{t-1} + \mathbf{b}^f\right)$$

$$\mathbf{o}_v^t = \sigma\left(\mathbf{W}^o \mathbf{x}_v^t + \sum_{l=1}^{K} \mathbf{U}_l^o \mathbf{h}_{vl}^{t-1} + \mathbf{b}^o\right) \tag{6.3}$$

$$\mathbf{u}_v^t = \tanh\left(\mathbf{W}^u \mathbf{x}_v^t + \sum_{l=1}^{K} \mathbf{U}_l^u \mathbf{h}_{vl}^{t-1} + \mathbf{b}^u\right)$$

$$\mathbf{c}_v^t = \mathbf{i}_v^t \odot \mathbf{u}_v^t + \sum_{l=1}^{K} \mathbf{f}_{vl}^t \odot \mathbf{c}_{vl}^{t-1}$$

$$\mathbf{h}_v^t = \mathbf{o}_v^t \odot \tanh(\mathbf{c}_v^t).$$

Compared to the Child-Sum Tree-LSTM, the N-ary Tree-LSTM introduces separate parameter matrices for each child k, which allows the model to learn more fine-grained representations for each node conditioned on the it's children.

6.3   GRAPH LSTM

The two types of Tree-LSTMs can be easily adapted to the graph. The graph-structured LSTM in Zayats and Ostendorf [2018] is an example of the N-ary Tree-LSTM applied to the graph. However, it is a simplified version since each node in the graph has at most two incoming edges (from its parent and sibling predecessor). Peng et al. [2017] propose another variant of the Graph LSTM based on the relation extraction task. The main difference between graphs and trees is that edges of graphs have their labels. And Peng et al. [2017] utilize different weight matrices to represent different labels:

$$\mathbf{i}_v^t = \sigma\left(\mathbf{W}^i \mathbf{x}_v^t + \sum_{k \in N_v} \mathbf{U}_{m(v,k)}^i \mathbf{h}_k^{t-1} + \mathbf{b}^i\right)$$

$$\mathbf{f}_{vk}^t = \sigma\left(\mathbf{W}^f \mathbf{x}_v^t + \mathbf{U}_{m(v,k)}^f \mathbf{h}_k^{t-1} + \mathbf{b}^f\right)$$

$$\mathbf{o}_v^t = \sigma\left(\mathbf{W}^o \mathbf{x}_v^t + \sum_{k \in N_v} \mathbf{U}_{m(v,k)}^o \mathbf{h}_k^{t-1} + \mathbf{b}^o\right) \tag{6.4}$$

$$\mathbf{u}_v^t = \tanh\left(\mathbf{W}^u \mathbf{x}_v^t + \sum_{k \in N_v} \mathbf{U}_{m(v,k)}^u \mathbf{h}_k^{t-1} + \mathbf{b}^u\right)$$

$$\mathbf{c}_v^t = \mathbf{i}_v^t \odot \mathbf{u}_v^t + \sum_{k \in N_v} \mathbf{f}_{vk}^t \odot \mathbf{c}_k^{t-1}$$

$$\mathbf{h}_v^t = \mathbf{o}_v^t \odot \tanh(\mathbf{c}_v^t),$$

where $m(v, k)$ denotes the edge label between node v and k, $\odot$ is the Hardamard product operation.

Liang et al. [2016] propose a Graph LSTM network to address the semantic object parsing task. It uses the confidence-driven scheme to adaptively select the starting node and determine the node updating sequence. It follows the same idea of generalizing the existing LSTMs into the graph-structured data but has a specific updating sequence while methods we mentioned above are agnostic to the order of nodes.

6.4 SENTENCE LSTM

Zhang et al. [2018c] propose the Sentence-LSTM (S-LSTM) for improving text encoding. It converts text into a graph and utilizes the Graph-LSTM to learn the representation. The S-LSTM shows strong representation power in many NLP problems.

In detail, the S-LSTM model regards each word as a node in the graph and it adds a supernode. For each layer, the word node could aggregate information from its adjacent words as well as the supernode. The supernode could aggregate information from all of the word nodes as well as itself. The connections of different nodes can be found in Figure 6.2.

The reason behind these settings of connections is that the supernode can provide global information to solve the long-distance dependency problem and the word node can model context information from its adjacent words. Thus, each word could obtain sufficient information and model both local and global information.

The S-LSTM model could be used in many natural language processing (NLP) tasks. The hidden states of words can be used to solve the word-level tasks such as sequence labeling, part-of-speech (POS) tagging and so on. The hidden state of the supernode can be used to solve sentence-level tasks such as sentence classification. The model has achieved promising results on several tasks and it also outperforms the powerful Transformer [Vaswani et al., 2017] model.

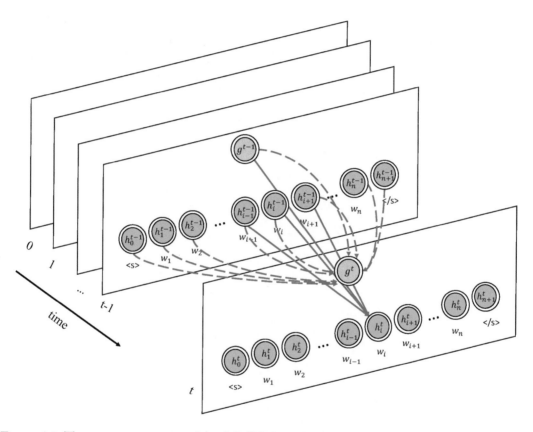

Figure 6.2: The propagation step of the S-LSTM model. The dash lines connect the supernode g with its neighbors from last layer. The solid lines connect the word node with its neighbors from last layer.

CHAPTER 7

Graph Attention Networks

The attention mechanism has been successfully used in many sequence-based tasks such as machine translation [Bahdanau et al., 2015, Gehring et al., 2017, Vaswani et al., 2017], machine reading [Cheng et al., 2016], and so on. Compared with GCN which treats all neighbors of a node equally, the attention mechanism could assign different attention score to each neighbor, thus identifying more important neighbors. It is intuitive to incorporate the attention mechanism into the propagation steps of Graph Neural Networks. In this chapter, we will talk about two variants: GAT and GAAN.

Note that graph attention networks can also be regarded as a method from the graph convolutional network family. We introduce this variant in a separate chapter to give more details.

7.1 GAT

Velickovic et al. [2018] propose a graph attention network (GAT) which incorporates the attention mechanism into the propagation steps. It follows the self-attention strategy and the hidden state of each node is computed by attending over its neighbors.

Velickovic et al. [2018] define a single *graph attentional layer* and constructs arbitrary graph attention networks by stacking this layer. The layer computes the coefficients in the attention mechanism of the node pair (i, j) by:

$$\alpha_{ij} = \frac{\exp\left(\text{LeakyReLU}\left(\mathbf{a}^T[\mathbf{W}\mathbf{h}_i \| \mathbf{W}\mathbf{h}_j]\right)\right)}{\sum_{k \in N_i} \exp\left(\text{LeakyReLU}\left(\mathbf{a}^T[\mathbf{W}\mathbf{h}_i \| \mathbf{W}\mathbf{h}_k]\right)\right)}, \tag{7.1}$$

where α_{ij} is the attention coefficient of node j to i, N_i represents the neighborhoods of node i in the graph. The input node features are denoted as $\mathbf{h} = \{\mathbf{h}_1, \mathbf{h}_2, \ldots, \mathbf{h}_N\}, \mathbf{h}_i \in \mathbb{R}^F$, where N is the number of nodes and F is the dimension, the output node features (with cardinality F') are denoted as $\mathbf{h}' = \{\mathbf{h}'_1, \mathbf{h}'_2, \ldots, \mathbf{h}'_N\}, \mathbf{h}'_i \in \mathbb{R}^{F'}$. $\mathbf{W} \in \mathbb{R}^{F' \times F}$ is the *weight matrix* of a shared linear transformation which applied to every node, $\mathbf{a} \in \mathbb{R}^{2F'}$ is the weight vector. It is normalized by a softmax function and the LeakyReLU nonlinearity (with negative input slop $\alpha = 0.2$) is applied.

Then the final output features of each node can be obtained by (after applying a nonlinearity σ):

$$\mathbf{h}'_i = \sigma\left(\sum_{j \in N_i} \alpha_{ij} \mathbf{W}\mathbf{h}_j\right). \tag{7.2}$$

Moreover, the layer utilizes the *multi-head attention* similarly to Vaswani et al. [2017] to stabilize the learning process. It applies K independent attention mechanisms to compute the

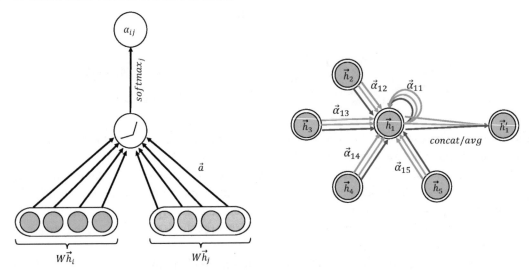

Figure 7.1: The illustration of the GAT model. **Left:** The attention mechanism employed in the model. **Right:** An illustration of multihead attention (with three heads denoted by different colors) by node 1 on its neighborhood.

hidden states and then concatenates their features (or computes the average), resulting in the following two output representations:

$$\mathbf{h}'_i = \overset{K}{\underset{k=1}{\Big\|}} \sigma\Big(\sum_{j \in N_i} \alpha_{ij}^k \mathbf{W}^k \mathbf{h}_j \Big) \tag{7.3}$$

$$\mathbf{h}'_i = \sigma\Big(\frac{1}{K} \sum_{k=1}^{K} \sum_{j \in N_i} \alpha_{ij}^k \mathbf{W}^k \mathbf{h}_j \Big), \tag{7.4}$$

where α_{ij}^k is normalized attention coefficient computed by the kth attention mechanism, $\|$ is the concatenation operation. The detailed model is illustrated in Figure 7.1.

The attention architecture in Velickovic et al. [2018] has several properties: (1) the computation of the node-neighbor pairs is parallelizable thus the operation is efficient; (2) it can deal with nodes with different degrees and assign corresponding weights to their neighbors; and (3) it can be applied to the inductive learning problems easily. As a result, GAT outperforms GCN in several tasks, such as semi-supervised node classification, link prediction, and so on.

7.2 GAAN

Besides GAT, Gated Attention Network (GaAN) [Zhang et al., 2018b] also uses the multi-head attention mechanism. The difference between the attention aggregator in GaAN and the one in GAT is that GaAN uses the key-value attention mechanism and the dot product attention while GAT uses a fully connected layer to compute the attention coefficients.

Furthermore, GaAN assigns different weights for different heads by computing an additional soft gate. This aggregator is called the gated attention aggregator. In detail, GaAN uses a convolutional network that takes the features of the center node and it neighbors to generate gate values. And as a result, it could outperform GAT as well as other GNN models with different aggregators in the inductive node classification problem.

CHAPTER 8

Graph Residual Networks

Many applications unroll or stack the graph neural network layer aiming to achieve better results as more layers (i.e., k layers) make each node aggregate more information from neighbors k hops away. However, it has been observed in many experiments that deeper models could not improve the performance and deeper models could even perform worse [Kipf and Welling, 2017]. This is mainly because more layers could also propagate the noisy information from an exponentially increasing number of expanded neighborhood members.

A straightforward method to address the problem, the residual network [He et al., 2016a], could be found from the computer vision community. But, even with residual connections, GCNs with more layers do not perform as well as the two-layer GCN on many datasets [Kipf and Welling, 2017]. In this chapter, we will talk about methods using skip connections to solve the problem and we call these models as GRNs.

8.1 HIGHWAY GCN

Rahimi et al. [2018] borrow ideas from the highway network [Zilly et al., 2016] and propose a Highway GCN which uses layer-wise gates. In each layer, the input is multiplied by gating weights and summed with the output ($\odot$ is the Hardamard product operation):

$$\begin{aligned} \mathbf{T}(\mathbf{h}^t) &= \sigma\left(\mathbf{W}^t\mathbf{h}^t + \mathbf{b}^t\right) \\ \mathbf{h}^{t+1} &= \mathbf{h}^{t+1} \odot \mathbf{T}\left(\mathbf{h}^t\right) + \mathbf{h}^t \odot \left(1 - \mathbf{T}\left(\mathbf{h}^t\right)\right). \end{aligned} \tag{8.1}$$

The aim of adding the highway gates is to provide the network the ability to select from new and old hidden states. Thus, an early hidden state could be propagated to the final state if it is needed. By adding the highway gates, the performance peaks at four layers and does not change much after adding more layers in a specific problem discussed in Rahimi et al. [2018].

The Column Network (CLN) proposed in Pham et al. [2017] also utilizes the highway network. But it has different function to compute the gating weights, which is selected based on the specific task.

8.2 JUMP KNOWLEDGE NETWORK

Xu et al. [2018] study the properties and limitations of neighborhood aggregation schemes. It shows that different nodes in graphs may need different receptive fields for better representation learning. For example, for a core node in a graph, the number of its neighbors may grow

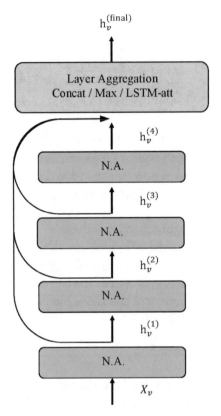

Figure 8.1: The illustration of Jump Knowledge Network. N.A. stands for neighborhood aggregation.

exponentially, thus much more noise may also be incorporated and the representation could be more smooth. For a node which is far from the core of the graph, the number of its neighbors could be very relatively small even if we expand its receptive field. Thus, this kind of nodes lacks sufficient information to learn better representations.

Xu et al. [2018] propose the Jump Knowledge Network which could learn adaptive, *structure-aware* representations. The Jump Knowledge Network selects from all of the intermediate representations (which "jump" to the last layer) for each node at the last layer, which enables each node to select effective neighborhood size as needed. Xu et al. [2018] used three approaches of **concatenation**, **max-pooling**, and **LSTM-attention** in the experiments to aggregate information. The illustration of JKN could be found in Figure 8.1.

The idea of Jump Knowledge Network is straightforward and it performs well on the experiments in social, bioinformatics and citation networks. It could also be combined with models like GCNs, GraphSAGE, and Graph Attention Networks to improve their performance.

8.3   DEEPGCNS

Li et al. [2019] borrow ideas from CNNs to add skip connections into graph neural networks. There are two major challenges to stack more layers of GNNs: vanishing gradient and over smoothing. Li et al. [2019] use residual connections and dense connections from ResNet [He et al., 2016b] and DenseNet [Huang et al., 2017] to solve the vanishing gradient problem and uses dilated convolutions [Yu and Koltun, 2015] to solve the over smoothing problem.

Li et al. [2019] denote the vanilla GCN as **PlainGCN** and further propose **ResGCN** and **DenseGCN**. In PlainGCN, the computation of hidden states is

$$\mathbf{H}^{t+1} = \mathcal{F}\left(\mathbf{H}^t, \mathbf{W}^t\right) \tag{8.2}$$

where $\mathcal{F}$ is a general graph convolution operation and $\mathbf{W}^t$ is the parameter at layer t.

For ResGCN, the computation can be denoted as

$$\begin{aligned}
\mathbf{H}^{t+1}_{Res} &= \mathbf{H}^{t+1} + \mathbf{H}^t \\
&= \mathcal{F}(\mathbf{H}^t, \mathbf{W}^t) + \mathbf{H}^t,
\end{aligned} \tag{8.3}$$

where the matrix of hidden states $\mathbf{H}^t$ is directly added to the matrix after the graph convolution.

For DenseGCN, the computation is

$$\begin{aligned}
\mathbf{H}^{t+1}_{Dense} &= \mathcal{T}\left(\mathbf{H}^{t+1}, \mathbf{H}^t, \ldots, \mathbf{H}^0\right) \\
&= \mathcal{T}\left(\mathcal{F}\left(\mathbf{H}^t, \mathbf{W}^t\right), \mathcal{F}\left(\mathbf{H}^{t-1}, \mathbf{W}^{t-1}\right), \ldots, \mathbf{H}^0\right),
\end{aligned} \tag{8.4}$$

where $\mathcal{T}$ is a vertex-wise concatenation function. Thus, the dimension of the hidden states grows with the graph layer. Figure 8.2 gives a straightforward demonstration of these three architectures.

Li et al. [2019] further use dilated convolutions [Yu and Koltun, 2015] to solve the over smoothing problem. The paper uses a *Dilated k-NN* method with dilation rate d. For each node, the method first computes the $k * d$ nearest neighbors using a pre-defined metric and then selects neighbors by skipping every d neighbors. For example, if $(u_1, u_2, \ldots, u_{k*d})$ are the $k * d$ nearest neighbors for a node v, then the dilated neighborhood of node v is $(u_1, u_{1+d}, u_{1+2d}, \ldots, u_{1+(k-1)d})$. The dilated convolution leverages information from different context and enlarges the receptive field of node v and is proven to be effective. Figure 8.3 shows the dilated convolution.

The dilated k-NN is added to the ResGCN and DenseGCN model. Li et al. [2019] conduct experiments on the task of point cloud semantic segmentation and built a 56-layer GCN to achieve promising results.

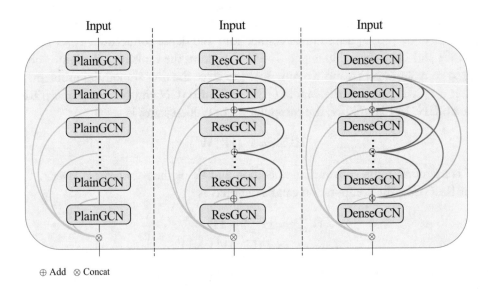

Figure 8.2: DeepGCN blocks (PlainGCN, ResGCN, DenseGCN) proposed in Li et al. [2019].

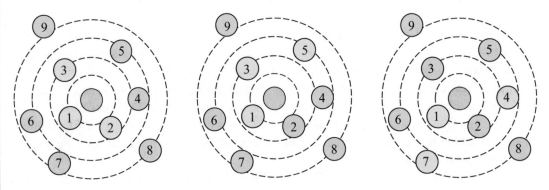

Figure 8.3: An example of the dilated convolution. The dilation rate is 1, 2, 3 for figures from left to right.

CHAPTER 9

Variants for Different Graph Types

The original GNN [Scarselli et al., 2009] works on the graphs that consist of nodes with label information and undirected edges, which is the simplest graph format. However, there are many variants of graphs in the world and modeling different graph types requires different GNN structures. In this chapter, we investigate graph neural networks designed for specific graph types.

9.1 DIRECTED GRAPHS

The first variant of graph is directed graph. Undirected edge which can be treated as two directed edges shows that there is a relation between two nodes. However, directed edges can bring more information than undirected edges. For example, in a knowledge graph where the edge starts from the head entity and ends at the tail entity, the head entity is the parent class of the tail entity, which suggests that we should treat the information propagation process from parent classes and child classes differently. Dense Graph Propagation (DGP) [Kampffmeyer et al., 2019] uses two kinds of weight matrix, $\mathbf{W}_a$ and $\mathbf{W}_d$, to incorporate more precise structural information. For each target node, it receives knowledge information from both all its descendants and its ancestors. The propagation rule is shown as follows:

$$H = \sigma \left(D_a^{-1} A_a \sigma \left(D_d^{-1} A_d X \Theta_d \right) \Theta_a \right), \tag{9.1}$$

where $D_a^{-1} A_a$, $D_d^{-1} A_d$ are the normalized adjacency matrix for parents and children, respectively. Since the various neighbors in the dense graph have different influence according to the distance, DGP proposed a weighting scheme that weighs a given node's neighbors in the dense graph. The authors use $w^a = \{w_i^a\}_{i=0}^K$ and $w^d = \{w_i^d\}_{i=0}^K$ to denote the weights for ancestors and descendants. The weighted propagation rule becomes

$$H = \sigma \left(\sum_{k=0}^K \alpha_k^a D_k^{a-1} A_k^a \sigma \left(\sum_{k=0}^K \alpha_k^d D_k^{d-1} A_k^d X \Theta_d \right) \Theta_a \right), \tag{9.2}$$

where A_k^a denotes the part of the adjacency matrice that contains the k-hop edges for the ancestor propagation phase and A_k^d contains the k-hop edges for the descendant propagation phase. D_k^a and D_k^d are the corresponding degree matrices for A_k^a and A_k^d.

9.2   HETEROGENEOUS GRAPHS

The second variant of graph is heterogeneous graph, where there are several kinds of nodes.

Definition 9.1   A **heterogeneous graph** can be represented as a directed graph $\mathcal{G} = \{\mathcal{V}, \mathcal{E}\}$ with a node type mapping $\phi : \mathcal{V} \to A$ and a relation type mapping $\psi : \mathcal{E} \to R$. $\mathcal{V}$ denotes the node set and $\mathcal{E}$ denotes the edge set. A denotes the node type set while R denotes the edge type set. $|A| > 1$ or $|R| > 1$ holds.

The simplest way to process heterogeneous graph is to convert the type of each node to a one-hot feature vector which is concatenated with the original feature. GraphInception [Zhang et al., 2018e] introduces the concept of metapath into the propagation on the heterogeneous graph.

Definition 9.2   A **meta-path** $\mathcal{P}$ of heterogeneous graph $\mathcal{G} = \{\mathcal{V}, \mathcal{E}\}$ is a path of the form $A_1 \xrightarrow{R_1} A_2 \xrightarrow{R_2} A_3 \cdots \xrightarrow{R_L} A_{L+1}$. $L + 1$ is the length of P.

With metapath, we can group the neighbors according to their node types and distances. In this way, the heterogeneous graph can be transformed to a set of homogeneous graphs and this is called *multi-channel network*. Formally, given a set of meta paths $\mathcal{S} = \{\mathcal{P}_1, \cdots, \mathcal{P}_{|\mathcal{S}|}\}$, the translated multi-channel network G' is defined as

$$G' = \left\{ G'_\ell \,\middle|\, G'_\ell = (\mathcal{V}_1, \mathcal{E}_{1\ell}), \ell = 1, \cdots, |\mathcal{S}| \right\}, \tag{9.3}$$

where $\mathcal{V}_1, \ldots, \mathcal{V}_m$ denote node sets with m different types, $\mathcal{V}_1$ is the target node type set, $\mathcal{E}_{1\ell} \subseteq \mathcal{V}_1 \times \mathcal{V}_1$ denotes the meta-path instances in $\mathcal{P}_\ell$. For each neighbor group, GraphInception treats it as a sub-graph in a homogeneous graph to do propagation and concatenates the propagation results from different homogeneous graphs to do a collective node representation. Instead of Laplacian matrix $\mathbf{L}$, GraphInception uses transition probability matrix $\mathbf{P}$ as the Fourier basis.

Recently, Wang et al. [2019b] propose the heterogeneous graph attention network (HAN) which utilizes node-level and semantic-level attentions. First for each meta-path, HAN learns a specific node embedding through node-level attention aggregation. Then with the meta-path specific embeddings, HAN presents semantic-level attention for a more comprehensive node embedding. Overall, the model have the ability to consider node importance and meta-paths importance simultaneously.

On event detection task, Peng et al. [2019] propose Pairwise Popularity Graph Convolutional Network (PP-GCN) to detect events on social networks. The model first calculates a weighted average between events for different meta-paths on event graphs. Then builds a weighted adjacent matrix to annotate social event instances and perform GCN [Kipf and Welling, 2017] on it to learn event embeddings.

In order to reduce training cost, ActiveHNE [Chen et al., 2019] introduces **active learning** into heterogeneous graph learning. Based on uncertainty and representativeness, Ac-

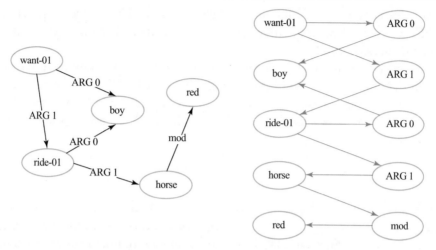

Figure 9.1: An example of AMR graph and its corresponding Levi graph. Left: The AMR graph of sentence *The boy wants to ride the red horse*. Right: The Levi transformation of the AMR graph.

tiveHNE selects the most valuable nodes in training set for label acquisition. This method significantly saves the query cost and achieves state-of-the-art performance on real-world datasets simultaneously.

9.3   GRAPHS WITH EDGE INFORMATION

In another variant of graph, each edge has additional information like the weight or the type of the edge. We list two ways to handle this kind of graphs: first, we can convert the graph to a bipartite graph where the original edges also become nodes and one original edge is split into two new edges which means there are two new edges between the edge node and begin/end nodes. This kind of graph transformation is called Levi graph transformation [Gross and Yellen, 2004, Levi, 1942]. Formally, given a graph $\mathcal{G} = \{\mathcal{V}, \mathcal{E}, L_{\mathcal{V}}, L_{\mathcal{E}}\}$, where $L_{\mathcal{V}}$ and $L_{\mathcal{E}}$ are labels of vertex set and edge set. Its corresponding Levi graph is defined as $\mathcal{G}' = \{\mathcal{V}', \mathcal{E}', L_{\mathcal{V}'}, L_{\mathcal{E}'}\}$, where $\mathcal{V}' = \mathcal{V} \cup \mathcal{E}$, $L_{\mathcal{V}'} = L_{\mathcal{V}} \cup L_{\mathcal{E}}$ and $L_{\mathcal{E}'} = \varnothing$. The new edge set $\mathcal{E}'$ contains edges from origin nodes and newly added edge nodes and edges in Levi graphs have no labels. In the work of G2S [Beck et al., 2018], the authors transform AMR graphs to their Levi graphs and apply gated graph neural networks. The encoder of G2S uses the following aggregation function for neighbors:

$$\mathbf{h}_v^t = \sigma \left(\frac{1}{|N_v|} \sum_{u \in N_v} \mathbf{W}_r \left(\mathbf{r}_v^t \odot \mathbf{h}_u^{t-1} \right) + \mathbf{b}_r \right), \tag{9.4}$$

where $\mathbf{r}_v^t$ is the reset gate in GRU for node v at layer t, $\mathbf{W}_r$ and $\mathbf{b}_r$ are the propagation parameters for different types of edges (relations), σ is the nonlinear activation function, and $\odot$ is the Hardamard product operation.

Second, we can adapt different weight matrices for the propagation on different kinds of edges. When the number of relations is very large, R-GCN [Schlichtkrull et al., 2018] introduces two kinds of regularization to reduce the number of parameters for modeling amounts of relations: *basis-* and *block-diagonal*-decomposition. With the basis decomposition, each $\mathbf{W}_r$ is defined as follows:

$$\mathbf{W}_r = \sum_{b=1}^{B} a_{rb} \mathbf{V}_b. \tag{9.5}$$

Here each $\mathbf{W}_r$ is a linear combination of basis transformations which can be viewed as some kind of weight sharing strategy. $\mathbf{V}_b \in \mathbb{R}^{d_{in} \times d_{out}}$ with coefficients a_{rb}. In the block-diagonal decomposition, R-GCN defines each $\mathbf{W}_r$ through the direct sum over a set of low-dimensional matrices, which needs more parameters than the first one:

$$\mathbf{W}_r = \bigoplus_{b=1}^{B} \mathbf{Q}_{br}. \tag{9.6}$$

Thereby, $\mathbf{W}_r = diag(\mathbf{Q}_{1r}, \ldots, \mathbf{Q}_{Br})$ is composed by $\mathbf{Q}_{br} \in \mathbb{R}^{(d^{(l+1)}/B) \times (d^{(l)}/B)}$. *Block-diagonal*-decomposition constrains the sparsity of weight matrices and encodes the hypothesis that latent vectors can be grouped into some small parts.

9.4   DYNAMIC GRAPHS

Spatial temporal forecasting is a crucial task which is widely applied in traffic forecasting, human action recognition, and climate prediction. Some of the forecasting tasks can be modeled as prediction tasks on dynamic graphs, which has static graph structure and dynamic input signals. As shown in Figure 9.2, given historical graph states, the goal is to predict the future graph states.

To capture both spatial and temporal information, DCRNN [Li et al., 2018d] and STGCN [Yu et al., 2018a] collect spatial and temporal information using independent modules. DCRNN [Li et al., 2018d] models the spatial graph flows as diffusion process on the graph. The diffusion convolution layers propagate spatial information and update nodes' hidden states. For temporal dependency, DCRNN leverages RNNs in which the matrix multiplication is replaced by diffusion convolution. The whole model is built under sequence-to-sequence architecture for multiple step forward forecasting. STGCN [Yu et al., 2018a] consists of multiple spatial-temporal convolutional blocks. In a spatial-temporal convolutional block, there are two temporal gated convolutional layers and one spatial graph convolutional layer between them. The residual connection and bottleneck strategy are adopted inside the blocks.

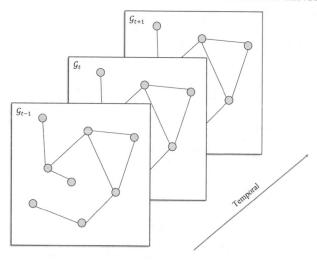

Figure 9.2: An example of spatial temporal graph. Each $\mathcal{G}_t$ indicates a frame of current graph state at time t.

Differently, Structural-RNN [Jain et al., 2016] and ST-GCN [Yan et al., 2018] collect spatial and temporal messages at the same time. They extend static graph structure with temporal connections so they can apply traditional GNNs on the extended graphs. Structural-RNN adds edges between the same node at time step t and $t + 1$ to construct the comprehension representation of spatio-temporal graphs. Then, the model represents each node with a nodeRNN and each edge with an edgeRNN. The edgeRNNs and nodeRNNs form a bipartite graph and forward-pass for each node.

ST-GCN [Yan et al., 2018] stacks graph frames of all time steps to construct spatial-temporal graphs. The model partitions the graph and assigns a weight vector for each node, then performs graph convolution directly on the weighted spatial-temporal graph.

Graph WaveNet [Wu et al., 2019d] considers a more challenging setting where the adjacency matrix of the static graph doesn't reflect the genuine spatial dependencies, i.e., some dependencies are missing or some are deceptive, which are ubiquitous because the distance of nodes doesn't necessarily mean a causal relationship. They propose a self-adaptive adjacency matrix which is learned in the framework and use a Temporal Convolution Network (TCN) together with a GCN to address the problem.

9.5   MULTI-DIMENSIONAL GRAPHS

So far we have considered the graph with binary edges. However, in the real world, nodes in a graph are connected by multiple relationship, forming a "multi-dimensional graph" (a.k.a. multi-view graph, multi-graph). For example, in the website YouTube for video sharing, interaction

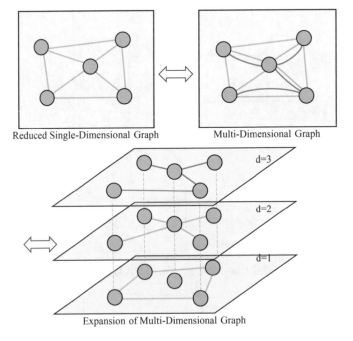

Figure 9.3: An example multi-dimensional graph, with its single graph and expansion.

type among users can be either "subscription," "sharing," and "comment" [Ma et al., 2019]. Because relation types are not assumed independent with each other naturally, directly applying the models of "single-dimensional" graph might not be an optimal solution.

Early works on multi-dimensional graph mainly focus on community detection and clustering. In Berlingerio et al. [2011], they give the illustration of "multidimensional community" and provide two different measures to disambiguate the definition of *density* in multi-dimensional graph. Papalexakis et al. [2013] provide concrete algorithms to find clusters across all the dimensions.

More recently, special types of GCN suitable for multi-dimensional graphs were designed. Ma et al. [2019] handle the problem by providing separate embeddings for a node in different dimensions, and these embeddings are viewed as projections from a general representation. They design an aggregation mechanism of graph neural network by taking into account the interactions between different nodes in the same dimension and the interactions between different dimensions of a single node. Khan and Blumenstock [2019] reduce the multi-dimensional graph to a single-dimensional one through two steps. They first merge the multiple view by subspace analysis and then prune the graph through manifold learning. The single-dimensional graph is passed into a normal GCN to perform learning. Sun et al. [2018] mainly focus on studying node embedding of the network and extends the node embedding algorithm (SVNE) to a multi-dimensional setting.

CHAPTER 10

Variants for Advanced Training Methods

As the original graph neural networks have some drawbacks in training and optimizing, in this section we introduce several variants with advanced training methods. We first introduce the sampling and the receptive filed control methods for effectiveness and scalability. Then we introduce several graph pooling methods. Finally, we introduce the data augmentation method and several unsupervised training methods.

10.1 SAMPLING

The original graph neural network has several drawbacks in training and optimization. For example, GCN requires the full-graph Laplacian, which is computational-consuming for large graphs. Furthermore, GCN is trained independently for a fixed graph, which lacks the ability for inductive learning.

GraphSAGE [Hamilton et al., 2017b] is a comprehensive improvement of the original GCN. To solve the problems mentioned above, GraphSAGE replaced full-graph Laplacian with learnable aggregation functions, which are key to perform message passing and generalize to unseen nodes. As shown in Eq. (5.20), they first aggregate neighborhood embeddings, concatenate with target node's embedding, then propagate to the next layer. With learned aggregation and propagation functions, GraphSAGE could generate embeddings for unseen nodes. Also, GraphSAGE uses a random neighbor sampling method to alleviate receptive field expansion.

Compared to GCN [Kipf and Welling, 2017], GraphSAGE proposes a way to train the model via batches of nodes instead of the full-graph Laplacian. This enables the training of large graphs though it may be time-consuming.

PinSage [Ying et al., 2018a] is an extension version of GraphSAGE on large graphs. It uses the importance-based sampling method. Simple random sampling is suboptimal because of the increase of variance. PinSage defines importance-based neighborhoods of node u as the T nodes that exert the most influence on node u. By simulating random walks starting from target nodes, this approach calculate the L_1-normalized visit count of nodes visited by the random walk. Then the top T nodes with the highest normalized visit counts with respect to u are selected to be the neighborhood of node u.

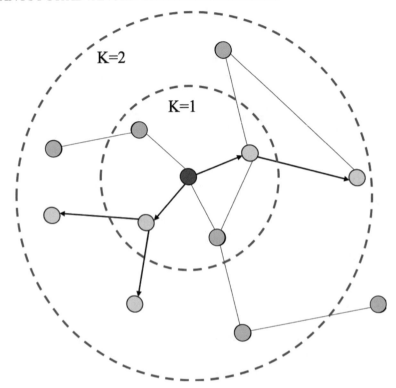

Figure 10.1: The illustration of sampled neighborhood on an example graph, K denotes the hop of neighborhood.

Opposed to node-wise sampling methods that should be performed independently for each node, layer-wise sampling only needs to be performed once. **FastGCN** [Chen et al., 2018a] further improves the sampling algorithm by interpreting graph convolution as integral transform of embedding functions under probability measure. Instead of sampling neighbors for each node, FastGCN directly samples the receptive field for each layer for variance reduction. FastGCN also incorporates importance sampling, in which the importance factor is calculated as below:

$$q(v) \propto \frac{1}{|N_v|} \sum_{u \in N_v} \frac{1}{|N_u|}, \tag{10.1}$$

where N_v is the neighborhood of node v. The sampling distribution is the same for each layer.

In contrast to fixed sampling methods above, Huang et al. [2018] introduce a parameterized and trainable sampler to perform layer-wise sampling. The authors try to learn a self-dependent function $g(x(u_j))$ of each node to determine its importance for sampling based on the node feature $x(u_j)$. The sampling distribution is defined as

$$q^*(u_j) = \frac{\sum_{i=1}^{n} p(u_j|v_i) |g(x(u_j))|}{\sum_{j=1}^{N} \sum_{i=1}^{n} p(u_j|v_i) |g(x(v_j))|}. \qquad (10.2)$$

Furthermore, this adaptive sampler could find optimal sampling importance and reduce variance simultaneously.

Many graph analytic problems are solved iteratively and finally achieve steady states. Following the idea of reinforcement learning, **SSE** [Dai et al., 2018] proposes Stochastic Fixed-Point Gradient Descent for GNN training to obtain the same steady-state solutions automatically from examples. This method views embedding update as value function and parameter update as policy function. In training, the algorithm samples nodes to update embeddings and samples labeled nodes to update parameters alternately.

Chen et al. [2018b] propose a control-variate based stochastic approximation algorithm for GCN by utilizing the historical activations of nodes as a control variate. This method maintains the historical average activations $\bar{h}_v^{(l)}$ to approximate the true activation $h_v^{(l)}$. The advantage of this approach is it limits the receptive field of nodes in the 1-hop neighborhood by using the historical hidden state as an affordable approximation, and the approximation are further proved to have zero variance.

10.2   HIERARCHICAL POOLING

In the area of computer vision, a convolutional layer is usually followed by a pooling layer to get more general features. Similar to these pooling layers, a lot of work focus on designing hierarchical pooling layers on graphs. Complicated and large-scale graphs usually carry rich hierarchical structures which are of great importance for node-level and graph-level classification tasks.

To explore such inner features, Edge-Conditioned Convolution (**ECC**) [Simonovsky and Komodakis, 2017] designs its pooling module with recursively downsampling operation. The downsampling method is based on splitting the graph into two components by the sign of the largest eigenvector of the Laplacian.

DIFFPOOL [Ying et al., 2018b] proposes a learnable hierarchical clustering module by training an assignment matrix in each layer:

$$\mathbf{S}^t = \text{softmax}\left(\text{GNN}_{pool}^l\left(\mathbf{A}^t, \mathbf{X}^t\right)\right), \qquad (10.3)$$

where $\mathbf{X}^t$ is the matrix of node features and $\mathbf{A}^t$ is the coarsened adjacency matrix of layer t.

10.3   DATA AUGMENTATION

Li et al. [2018a] focus on the limitations of GCN, which include that GCN requires many additional labeled data for validation and also suffers from the localized nature of the convolutional filter. To solve the limitations, the authors propose Co-Training GCN and Self-Training GCN to enlarge the training dataset. The former method finds the nearest neighbors of training data while the latter one follows a boosting-like way.

10.4   UNSUPERVISED TRAINING

GNNs are typically used for supervised or semi-supervised learning problems. Recently, there has been a trend to extend auto-encoder (AE) to graph domains. Graph auto-encoders aim at representing nodes into low-dimensional vectors by an unsupervised training manner.

Graph Auto-Encoder (GAE) [Kipf and Welling, 2016] first uses GCNs to encode nodes in the graph. Then it uses a simple decoder to reconstruct the adjacency matrix and computes the loss from the similarity between the original adjacency matrix and the reconstructed matrix (σ is the nonlinear activation function):

$$\mathbf{Z} = \text{GCN}(\mathbf{X}, \mathbf{A})$$
$$\widetilde{A} = \sigma \left(\mathbf{Z}\mathbf{Z}^T \right). \tag{10.4}$$

Kipf and Welling [2016] also train the GAE model in a variational manner and the model is named as the variational graph auto-encoder (VGAE). Furthermore, Berg et al. use GAE in recommender systems and have proposed the graph convolutional matrix completion model (GC-MC) [van den Berg et al., 2017], which outperforms other baseline models on the Movie-Lens dataset.

Adversarially Regularized Graph Auto-encoder (ARGA) [Pan et al., 2018] employs generative adversarial networks (GANs) to regularize a GCN-based graph auto-encoder to follow a prior distribution.

Deep Graph Infomax (DGI) [Veličković et al., 2019] aims to maximize the local-global mutual information to learn representations. The local information comes from each node's hidden state after the graph convolution function $\mathcal{F}$. The global information $\vec{s}$ of a graph is computed by the readout function $\mathcal{R}$. This function aggregates all node presentations and is set to an average function in the paper. The paper uses node shuffling to get negative examples (by changing node features from $\mathbf{X}$ to $\tilde{\mathbf{X}}$ with a corruption function $\mathcal{C}$). Then it use a discriminator $\mathcal{D}$ to classify the positive samples and negative samples. The architecture of DGI is shown in Figure 10.2.

There are also several graph auto-encoders such as NetRA [Yu et al., 2018b], DNGR [Cao et al., 2016], SDNE [Wang et al., 2016], and DRNE [Tu et al., 2018], however, they don't use GNNs in their framework.

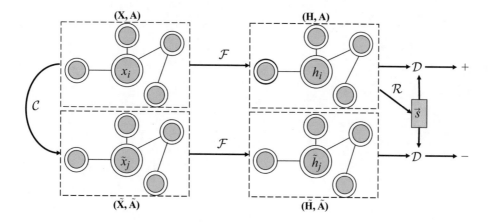

Figure 10.2: The architecture of Deep Graph Infomax.

CHAPTER 11

General Frameworks

Apart from different variants of graph neural networks, several general frameworks are proposed aiming to integrate different models into one single framework. Gilmer et al. [2017] propose the message passing neural network (MPNN) and it is a unified framework to generalize several graph neural network and graph convolutional network methods. Wang et al. [2018b] propose the non-local neural network (NLNN) which is used to solve computer vision tasks. It could generalize several "self-attention"-style methods [Hoshen, 2017, Vaswani et al., 2017, Velickovic et al., 2018]. Battaglia et al. [2018] propose the graph network (GN) which unified the MPNN and NLNN methods as well as many other variants like Interaction Networks [Battaglia et al., 2016, Watters et al., 2017], Neural Physics Engine [Chang et al., 2017], CommNet [Sukhbaatar et al., 2016], structure2vec [Dai et al., 2016, Khalil et al., 2017], GGNN [Li et al., 2016], Relation Network [Raposo et al., 2017, Santoro et al., 2017], Deep Sets [Zaheer et al., 2017], and Point Net [Qi et al., 2017a].

11.1 MESSAGE PASSING NEURAL NETWORKS

Gilmer et al. [2017] propose a general framework for supervised learning on graphs called message passing neural networks (MPNNs). The MPNN provides a unified framework by considering the commonalities among several popular graph models such as spectral approaches [Bruna et al., 2014, Defferrard et al., 2016, Kipf and Welling, 2017] and non-spectral approaches [Duvenaud et al., 2015] in graph convolution, gated graph neural networks [Li et al., 2016], interaction networks [Battaglia et al., 2016], molecular graph convolutions [Kearnes et al., 2016], deep tensor neural networks [Schütt et al., 2017], and so on.

The model contains two phases, a message passing phase and a readout phase. The message passing phase (namely, the propagation step) runs for T time steps and contains two subfunctions: a message function M_t and a vertex update function U_t. Using messages $\mathbf{m}_v^t$, the updating functions of hidden states $\mathbf{h}_v^t$ are as follows:

$$
\begin{aligned}
\mathbf{m}_v^{t+1} &= \sum_{w \in N_v} M_t \left(\mathbf{h}_v^t, \mathbf{h}_w^t, \mathbf{e}_{vw} \right) \\
\mathbf{h}_v^{t+1} &= U_t \left(\mathbf{h}_v^t, \mathbf{m}_v^{t+1} \right),
\end{aligned}
\tag{11.1}
$$

where $\mathbf{e}_{vw}$ represents features of the edge from node v to w. The readout phase uses a readout function R to compute a representation for the whole graph

$$
\hat{\mathbf{y}} = R \left(\{ \mathbf{h}_v^T | v \in G \} \right),
\tag{11.2}
$$

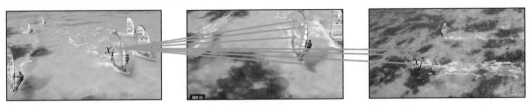

Figure 11.1: A spacetime non-local operation in the network trained for video classification. The response of x_i is computed as the weighted sum of all positions x_j where in this figure only the highest weighted ones are shown.

where T denotes the total time steps. The message function M_t, vertex update function U_t, and readout function R could have different settings. Hence, the MPNN framework could generalize several different models via different function settings. Here we give an example of generalizing GGNN, and other models' function settings could be found in Gilmer et al. [2017]. The function settings for GGNNs are:

$$
\begin{aligned}
M_t \left(\mathbf{h}_v^t, \mathbf{h}_w^t, \mathbf{e}_{vw} \right) &= \mathbf{A}_{\mathbf{e}_{vw}} \mathbf{h}_w^t \\
U_t &= GRU \left(\mathbf{h}_v^t, \mathbf{m}_v^{t+1} \right) \\
R &= \sum_{v \in V} \sigma \left(i \left(\mathbf{h}_v^T, \mathbf{h}_v^0 \right) \right) \odot \left(j \left(\mathbf{h}_v^T \right) \right),
\end{aligned}
\tag{11.3}
$$

where $\mathbf{A}_{\mathbf{e}_{vw}}$ is the adjacency matrix, one for each edge label e. GRU is the Gated Recurrent Unit introduced in Cho et al. [2014]. i and j are neural networks in function R.

11.2   NON-LOCAL NEURAL NETWORKS

Wang et al. [2018b] propose the Non-local Neural Networks (NLNN) for capturing long-range dependencies with deep neural networks (DNNs). The non-local operation is a generalization of the classical non-local mean operation [Buades et al., 2005] in computer vision. The non-local operation computes the weighted sum of the features at all positions for a specific position. The set of positions can come from both the time dimension and the space dimension. An example of NLNN used on the video classification task could be found in Figure 11.1.

The NLNN can be viewed as a unification of different "self-attention"-style methods [Hoshen, 2017, Vaswani et al., 2017, Velickovic et al., 2018]. We will first introduce the general definition of non-local operations and then some specific instantiations.

Following the non-local mean operation [Buades et al., 2005], the generic non-local operation is defined as:

$$
\mathbf{h}_i' = \frac{1}{\mathcal{C}(\mathbf{h})} \sum_{\forall j} f \left(\mathbf{h}_i, \mathbf{h}_j \right) g \left(\mathbf{h}_j \right),
\tag{11.4}
$$

where i is the target position and the selection of j should enumerate all possible positions. $f(\mathbf{h}_i, \mathbf{h}_j)$ is used to compute the "attention" between position i and j. $g(\mathbf{h}_j)$ denotes a transformation of the input $\mathbf{h}_j$ and a factor $\frac{1}{C(\mathbf{h})}$ is utilized to normalize the results.

There are several instantiations with different f and g settings. For simplicity, Wang et al. [2018b] use the linear transformation as the function g. That means $g(\mathbf{h}_j) = \mathbf{W}_g \mathbf{h}_j$, where $\mathbf{W}_g$ is a learned weight matrix. Next, we list the choices for function f in the following.

Gaussian. The Gaussian function is a natural choice according to the non-local mean [Buades et al., 2005] and bilateral filters [Tomasi and Manduchi, 1998]. Thus:

$$f\left(\mathbf{h}_i, \mathbf{h}_j\right) = e^{\mathbf{h}_i^T \mathbf{h}_j}, \tag{11.5}$$

here $\mathbf{h}_i^T \mathbf{h}_j$ is dot-product similarity and $C(\mathbf{h}) = \sum_{\forall j} f(\mathbf{h}_i, \mathbf{h}_j)$.

Embedded Gaussian. It is straightforward to extend the Gaussian function by computing similarity in the embedding space, which means:

$$f\left(\mathbf{h}_i, \mathbf{h}_j\right) = e^{\theta(\mathbf{h}_i)^T \phi(\mathbf{h}_j)}, \tag{11.6}$$

where $\theta(\mathbf{h}_i) = \mathbf{W}_\theta \mathbf{h}_i$, $\phi(\mathbf{h}_j) = W_\phi \mathbf{h}_j$ and $C(\mathbf{h}) = \sum_{\forall j} f(\mathbf{h}_i, \mathbf{h}_j)$.

It could be found that the self-attention proposed in Vaswani et al. [2017] is a special case of the Embedded Gaussian version. For a given i, $\frac{1}{C(\mathbf{h})} f(\mathbf{h}_i, \mathbf{h}_j)$ becomes the *softmax* computation along dimension j. So that $\mathbf{h}' = \text{softmax}(\mathbf{h}^T \mathbf{W}_\theta^T \mathbf{W}_\phi \mathbf{h}) g(\mathbf{h})$, which matches the form of self-attention in Vaswani et al. [2017].

Dot product. The function f can also be implemented as dot-product similarity:

$$f\left(\mathbf{h}_i, \mathbf{h}_j\right) = \theta\left(\mathbf{h}_i\right)^T \phi\left(\mathbf{h}_j\right). \tag{11.7}$$

Here the factor $C(\mathbf{h}) = N$, where N is the number of positions in $\mathbf{h}$.

Concatenation. Here we have:

$$f\left(\mathbf{h}_i, \mathbf{h}_j\right) = \text{ReLU}\left(\mathbf{w}_f^T \left[\theta\left(\mathbf{h}_i\right) \| \phi\left(\mathbf{h}_j\right)\right]\right), \tag{11.8}$$

where $\mathbf{w}_f$ is a weight vector projecting the vector to a scalar and $C(\mathbf{h}) = N$.

Wang et al. [2018b] further propose a non-local block by using the non-local operation mentioned above:

$$\mathbf{z}_i = \mathbf{W}_z \mathbf{h}_i' + \mathbf{h}_i, \tag{11.9}$$

where $\mathbf{h}_i'$ is given in Eq. (11.4) and "$+\mathbf{h}_i$" denotes the residual connection [He et al., 2016a]. Hence, the non-local block could be insert into any pre-trained model, which makes the block more applicable. Wang et al. [2018b] conduct experiments on the tasks of video classification, object detection and segmentation, and pose estimation. And on these tasks, the simple addition of non-local blocks leads to a significant improvement over baselines.

11.3 GRAPH NETWORKS

Battaglia et al. [2018] propose the GN framework which generalizes and extends various graph neural network, MPNN and NLNN approaches [Gilmer et al., 2017, Scarselli et al., 2009, Wang et al., 2018b]. We first introduce the graph definition in Battaglia et al. [2018], then we describe the GN block, a core GN computation unit, and its computational steps, and finally we will introduce its basic design principles.

Graph definition. In Battaglia et al. [2018], a graph is defined as a 3-tuple $G = (\mathbf{u}, H, E)$ (here we use H instead of V for notation's consistency). $\mathbf{u}$ is a global attribute, $H = \{\mathbf{h}_i\}_{i=1:N^v}$ is the node set (with dimension N^v), where each $\mathbf{h}_i$ denotes the feature of the node. $E = \{(\mathbf{e}_k, r_k, s_k)\}_{k=1:N^e}$ is the edge set (with dimension N^e), where each $\mathbf{e}_k$ denotes the feature of the edge, r_k denotes the receiver node and s_k denotes the sender node.

GN block. A GN block contains three "update" functions, ϕ, and three "aggregation" functions, ρ,

$$
\begin{aligned}
\mathbf{e}'_k &= \phi^e\left(\mathbf{e}_k, \mathbf{h}_{r_k}, \mathbf{h}_{s_k}, \mathbf{u}\right) & \bar{\mathbf{e}}'_i &= \rho^{e \to h}\left(E'_i\right) \\
\mathbf{h}'_i &= \phi^h\left(\bar{\mathbf{e}}'_i, \mathbf{h}_i, \mathbf{u}\right) & \bar{\mathbf{e}}' &= \rho^{e \to u}\left(E'\right) \qquad (11.10) \\
\mathbf{u}' &= \phi^u\left(\bar{\mathbf{e}}', \bar{\mathbf{h}}', \mathbf{u}\right) & \bar{\mathbf{h}}' &= \rho^{h \to u}\left(H'\right),
\end{aligned}
$$

where $\quad E'_i = \{(\mathbf{e}'_k, r_k, s_k)\}_{r_k=i,\, k=1:N^e}, \quad H' = \{\mathbf{h}'_i\}_{i=1:N^v}, \quad$ and $\quad E' = \bigcup_i E'_i = \{(\mathbf{e}'_k, r_k, s_k)\}_{k=1:N^e}$. The design of the ρ functions should consider the number and the order of the inputs. The results of these functions must be invariant to these factors.

Computation steps. The computation steps of a GN block are as follows.

1. ϕ^e is applied per edge. The result set of each edge for node i is denoted as $E'_i = \{(\mathbf{e}'_k, r_k, s_k)\}_{r_k=i,\, k=1:N^e}$. And $E' = \bigcup_i E'_i = \{(\mathbf{e}'_k, r_k, s_k)\}_{k=1:N^e}$ is the set of all outputs of the edges.

2. $\rho^{e \to h}$ uses E'_i to aggregate corresponding edge updates for node i and get the result $\bar{\mathbf{e}}'_i$.

3. ϕ^h is used to compute the updated node representation $\mathbf{h}'_i$ for node i. The set of all nodes' representations is $H' = \{\mathbf{h}'_i\}_{i=1:N^v}$.

4. $\rho^{e \to u}$ uses E' to aggregate all edge updates into $\bar{\mathbf{e}}'$. It will be further used in the computation of the global state.

5. $\rho^{h \to u}$ uses H' to aggregate all node updates into $\bar{\mathbf{h}}'$, which will be used in the update of the global state.

6. ϕ^u is designed to compute an update for the global attribute $\mathbf{u}'$ with the information from $\bar{\mathbf{e}}'$, $\bar{\mathbf{h}}'$ and $\mathbf{u}$.

Note here the order is not strictly enforced. For example, it is possible to proceed from global, to per-node, to per-edge updates. And the ϕ and ρ functions need not be neural networks though in this paper we only focus on neural network implementations.

Design Principles. The design of GN based on three basic principles: flexible representations, configurable within-block structure, and composable multi-block architectures.

- **Flexible representations.** The GN framework supports flexible representations of the attributes as well as different graph structures. The global, node, and edge attributes can use different kinds of representations and researchers usually use real-valued vectors and tensors. One can simply tailor the output of a GN block according to specific demands of tasks. For example, Battaglia et al. [2018] list several *edge-focused* [Hamrick et al., 2018, Kipf et al., 2018], *node-focused* [Battaglia et al., 2016, Chang et al., 2017, Sanchez et al., 2018, Wang et al., 2018a], and *graph-focused* [Battaglia et al., 2016, Gilmer et al., 2017, Santoro et al., 2017] GNs. In terms of graph structures, the framework can be applied to both structural scenarios where the graph structure is explicit and non-structural scenarios where the relational structure should be inferred or assumed.

- **Configurable within-block structure.** The functions and their inputs within a GN block can have different settings so that the GN framework provides flexibility in within-block structure configuration. For example, Hamrick et al. [2018] and Sanchez et al. [2018] use the full GN blocks. Their ϕ implementations use neural networks and their ρ functions use the elementwise summation. Based on different structure and functions settings, a variety of models (such as MPNN, NLNN, and other variants) could be expressed by the GN framework. Figure 11.2a gives an illustration of a full GN block and other models can be regarded as special variants of the GN block. For example, the MPNN uses the features of nodes and edges as input and outputs graph-level and node-level representations. The MPNN model does not use the graph-level input features and omits the learning process of edge embeddings.

- **Composable multi-block architectures.** GN blocks could be composed to construct complex architectures. Arbitrary numbers of GN blocks could be composed in sequence with shared or unshared parameters. Battaglia et al. [2018] utilize GN blocks to construct an *encode-process-decode* architecture and a recurrent GN-based architecture. These architectures are demonstrated in Figure 11.3. Other techniques for building GN-based architectures could also be useful, such as skip connections, LSTM-, or GRU-style gating schemes and so on.

In conclusion, GN is a general and flexible framework for deep learning on graphs. It can be used for various tasks including physical systems, traffic networks and so on. However, the GNs still has its limitations. For example, it cannot solve some classes of problems like discriminating between certain non-isomorphic graphs.

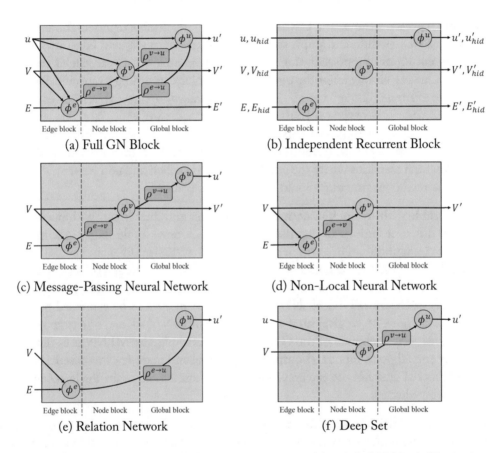

Figure 11.2: Different internal GN block configurations. (a) a full GN block [Battaglia et al., 2018]; (b) an independent recurrent block [Sanchez et al., 2018]; (c) an MPNN [Gilmer et al., 2017]; (d) a NLNN [Wang et al., 2018b]; (e) a relation network [Raposo et al., 2017]; and (f) a deep set [Zaheer et al., 2017].

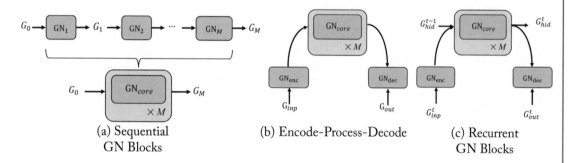

Figure 11.3: Examples of architectures composed by GN blocks. (a) The sequential processing architecture; (b) The encode-process-decode architecture; and (c) The recurrent architecture.

CHAPTER 12

Applications – Structural Scenarios

In the following sections, we will introduce GNN's applications in structural scenarios, where the data are naturally performed in the graph structure. For example, GNNs are widely being used in social network prediction [Hamilton et al., 2017b, Kipf and Welling, 2017], traffic prediction [Rahimi et al., 2018], recommender systems [van den Berg et al., 2017, Ying et al., 2018a], and graph representation [Ying et al., 2018b]. Specifically, we are discussing how to model real-world physical systems with object-relationship graphs, how to predict chemical properties of molecules and biological interaction properties of proteins and the methods of reasoning about the out-of-knowledge-base (OOKB) entities in knowledge graphs.

12.1 PHYSICS

Modeling real-world physical systems is one of the most basic aspects of understanding human intelligence. By representing objects as nodes and relations as edges, we can perform GNN-based reasoning about objects, relations, and physics in a simplified but effective way.

Battaglia et al. [2016] propose *Interaction Networks* to make predictions and inferences about various physical systems. In current state, we input objects and relations into GNN to model their interactions, then the physical dynamics are adopted to predict future states. They separately model relation-centric and object-centric models, making it easier to generalize across different systems.

In CommNet [Sukhbaatar et al., 2016], interactions are not modeled explicitly. Instead, an interaction vector is obtained by averaging all other agents' hidden vectors.

VAIN [Hoshen, 2017] further introduces attentional methods into agent interaction process, which preserves both the complexity advantages and computational efficiency as well.

Visual Interaction Networks [Watters et al., 2017] could make predictions from pixels. It learns a state code from two consecutive input frames for each object. Then, after adding their interaction effect by an Interaction Net block, the state decoder converts state codes to next step's state.

Sanchez et al. [2018] propose a GN-based model which could either perform state prediction or inductive inference. The inference model takes partially observed information as input and constructs a hidden graph for implicit system classification. Kipf et al. [2018] also build graphs from object trajectories, they adopt an encoder-decoder architecture for neural relational

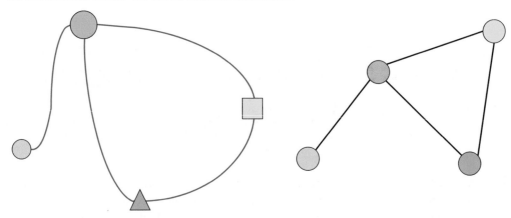

Figure 12.1: A physical system and its corresponding graph representation. Colored nodes denote different objects and edges denote interaction between them.

inference process. In detail, the encoder returns a factorized distribution of interaction graph $\mathcal{Z}$ through GNN while the decoder generates trajectory predictions conditioned on both the latent code of the encoder and the previous time step of the trajectory.

On the problem of solving partial differential equations, inspired by finite element methods [Hughes, 2012], *graph element networks* [Alet et al., 2019] place nodes in the continuous space. Each node represents the local state of the system, and the model establishes a connectivity graph among the nodes. GNN propagates state information to simulate the dynamic system.

12.2   CHEMISTRY AND BIOLOGY

Molecules and proteins are structured entities that can be represented by graphs. As shown in Figure 12.2, atoms or residues are nodes and chemical bonds or chains are edges. By GNN-based representation learning, the learned vectors can help with drug design, chemical reaction prediction and interaction prediction.

12.2.1   MOLECULAR FINGERPRINTS

Molecular fingerprints are features vectors representing molecules, which play a key role in computer-aided drug design. Traditional molecular fingerprints rely on heuristic methods which are hand-crafted. GNN provides more flexible approaches for better fingerprints. Duvenaud et al. [2015] propose *neural graph fingerprints* which calculate substructure feature vectors via GCN and sum to get overall representation. The aggregation function is

$$\mathbf{h}^t_{N_v} = \sum_{u \in N_v} \text{CONCAT}\left(\mathbf{h}^t_u, \mathbf{e}_{uv}\right), \tag{12.1}$$

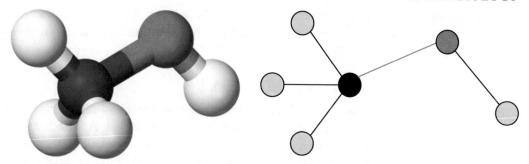

Figure 12.2: A single CH_3OH molecular and its graph representation. Nodes are elements and edges are bonds.

where $\mathbf{e}_{uv}$ is the edge feature of edge (u, v). Then update node representation by

$$\mathbf{h}_v^{t+1} = \sigma \left(\mathbf{W}_t^{\deg(v)} \mathbf{h}_{N_v}^t \right), \tag{12.2}$$

where $\deg(v)$ is the degree of node v and $\mathbf{W}_t^N$ is a learned matrix for each time step t and node degree N.

Kearnes et al. [2016] further explicitly model atom and atom pairs independently to emphasize atom interactions. It introduces edge representation $\mathbf{e}_{uv}^t$ instead of aggregation function, i.e., $\mathbf{h}_{N_v}^t = \sum_{u \in N_v} \mathbf{e}_{uv}^t$. The node update function is

$$\mathbf{h}_v^{t+1} = \text{ReLU} \left(\mathbf{W}_1 \left(\text{ReLU} \left(\mathbf{W}_0 \mathbf{h}_u^t \right), \mathbf{h}_{N_v}^t \right) \right) \tag{12.3}$$

while the edge update function is

$$\mathbf{e}_{uv}^{t+1} = \text{ReLU} \left(\mathbf{W}_4 \left(\text{ReLU} \left(\mathbf{W}_2 \mathbf{e}_{uv}^t \right), \text{ReLU} \left(\mathbf{W}_3 \left(\mathbf{h}_v^t, \mathbf{h}_u^t \right) \right) \right) \right). \tag{12.4}$$

Beyond atom molecular graphs, some works [Jin et al., 2018, 2019] represent molecules as junction trees. A junction tree is generated by contracting certain vertices in corresponding molecular graph into a single node. The nodes in a junction tree are molecular substructures such as rings and bonds. Jin et al. [2018] leverage variational auto-encoder to generate molecular graphs. Their model follows a two-step process, first generating a junction tree scaffold over chemical substructures, then combining them into a molecule with a graph message passing network. Jin et al. [2019] focus on molecular optimization. This task aims to map one molecule to another molecular graph which preserves better properties. The proposed VJTNN uses graph attention to decode the junction tree and incorporates GAN for adversarial training to avoid valid graph translation.

To better explain the function of each substructure in a molecule, Lee et al. [2019] propose a game-theoretic approach to exhibit the transparency in structured data. The model is set up as a two-player co-operative game between a *predictor* and a *witness*. The *predictor* is trained to minimize the discrepancy while the goal of the *witness* is to test how well the *predictor* conforms to the transparency.

12.2.2   CHEMICAL REACTION PREDICTION

Chemical reaction product prediction is a fundamental problem in organic chemistry. Do et al. [2019] view chemical reaction as graph transformation process and introduces GTPN model. GTPN uses GNN to learn representation vectors of reactant and reagent molecules, then leverages reinforcement learning to predict the optimal reaction path in the form of bond change which transforms the reactants into products. Bradshaw et al. [2019] give another view that chemical reactions can be described as the stepwise redistribution of electrons in molecules. Their model tries to predict the electron paths by learning path distribution over the electron movements. They represent node and graph embeddings with a four-layer GGNN, and then optimize the factorized path generation probability.

12.2.3   MEDICATION RECOMMENDATION

Using deep learning algorithms to help recommend medications has been explored by researchers and doctors extensively. The traditional methods can be categorized into instance-based and longitudinal electronic health records (EHR)-based medication recommendation methods.

To fill the gap between them, Shang et al. [2019c] propose GAMENet which takes both longitudinal patient EHR data and drug knowledge based on drug-drug interactions (DDI) as inputs. GAMENet embeds both EHR graph and DDI graph, then feed them into Memory Bank for final output.

To further exploit the hierarchical knowledge for meditation recommendation, Shang et al. [2019b] combine the power of GNN and BERT for medical code representation. The authors first encode the internal hierarchical structure with GNN, and then feed the embeddings into the pre-trained EHR encoder and the fine-tuned classifier for downstream predictive tasks.

12.2.4   PROTEIN AND MOLECULAR INTERACTION PREDICTION

Fout et al. [2017] focus on the task named protein interface prediction, which is a challenging problem to predict the interaction between proteins and the interfaces they occur. The proposed GCN-based method, respectively, learns ligand and receptor protein residue representation and merges them for pairwise classification. Xu et al. [2019b] introduce MR-GNN which utilizes a multi-resolution model to capture multi-scale node features. The model also utilizes two long short-term memory networks to capture the interaction between two graphs step-by-step.

GNN can also be used in biomedical engineering. With Protein-Protein Interaction Network, Rhee et al. [2018] leverage graph convolution and relation network for breast cancer subtype classification. Zitnik et al. [2018] also suggest a GCN-based model for polypharmacy side

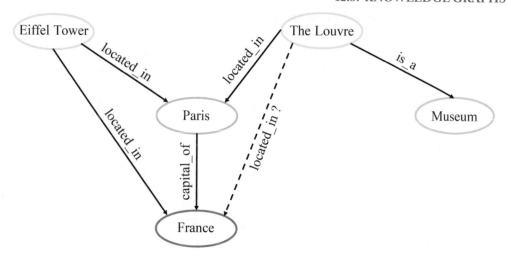

Figure 12.3: Example of knowledge base fragment. The nodes are *entities* and the edges are *relations*. The dashed line is missing edge information to be inferred.

effects prediction. Their work models the drug and protein interaction network and separately deals with edges in different types.

12.3   KNOWLEDGE GRAPHS

Knowledge graphs (KGs) represent knowledge bases (KBs) as a directed graph whose nodes and edges represent *entities* and *relations* between entities, respectively. The relationships are organized in the forms of *(head, relation, tail)* (denoted *(h,r,t)*) triplets. These KGs are widely used in recommendation, web search, and question answering.

12.3.1   KNOWLEDGE GRAPH COMPLETION

To effectively encode knowledge graphs into a low-dimensional continuous vector space, GNN has been a widely used efficient tool for incorporating the topological structure of knowledge graphs. Link prediction and entity classification are two major tasks in KB completion.

R-GCN proposed by Schlichtkrull et al. [2018] is the first GNN-based framework for modeling relational data with parameter sharing techniques to enforce sparsity constraints. They also proved combining conventional factorization model like DistMult [Yang et al., 2015a] with GCN structure as decoder achieved better performance on standard link prediction benchmarks.

Shang et al. [2019a] take the benefits of GNN and *ConvE* [Dettmers et al., 2018] together and introduces a novel end-to-end Structure-Aware Convolutional Network (SACN). SACN consists of a GCN-based encoder and a CNN-based decoder. The encoder uses a stack of GCN layers to learn entity and relation embeddings while the decoder feeds the embeddings into multi-channel CNN for vectorization and projection, the output vectors are matched with all candidates by inner product.

Nathani et al. [2019] apply GATs as encoders to capture the diversity of roles played by an entity in various relations. Furthermore, to alleviate the contribution decrease in message propagation process, the authors introduce auxiliary edges between multi-hop neighbors, which allow the direct flow of knowledge between entities.

12.3.2   INDUCTIVE KNOWLEDGE GRAPH EMBEDDING

The inductive knowledge graph embedding aims to answer queries concerning out-of-knowledge-base entities, which are test entities that are not observed at training time.

Hamaguchi et al. [2017] utilize GNNs to solve the out-of-knowledge-base (OOKB) entity problem in knowledge base completion (KBC). The OOKB entities in Hamaguchi et al. [2017] are directly connected to the existing entities thus the embeddings of OOKB entities can be aggregated from the existing entities. The method achieves satisfying performance both in the standard KBC setting and the OOKB setting.

For finer aggregation process, Wang et al. [2019a] designs a attention aggregator to learn the embeddings of OOKB entities. The attention weights have two parts: statistical logic rule mechanism to measure the neighboring relations' usefulness and neural network mechanism to measure the importance of neighboring nodes.

12.3.3   KNOWLEDGE GRAPH ALIGNMENT

Knowledge graphs encode rich knowledge in single language or domain but lack cross-lingual or cross-domain links to bridge the gap. Therefore, the knowledge graph alignment task is proposed to solve the problem.

Wang et al. [2018f] use GCNs to solve the cross-lingual knowledge graph alignment problem. The model embeds entities from different languages into a unified embedding space and aligns them based on the embedding similarity.

To better leverage context information, Xu et al. [2019a] propose *topic entity graph* to represent the KG contextual environment of an entity. The *topic entity graph* consists of the target entity and its 1-hop neighborhood. For alignment task, Xu et al. [2019a] use a graph matching network to match the two *topic entity graphs*, and propagates the local matching information by another GCN.

Focusing on linking two *large-scale heterogeneous academic entity graphs*, Zhang et al. [2019] adopt three specific modules to align venues, papers and authors. The venue linking module is based on LSTM to deal with name sequences, the paper-linking module consists of local sensitive hashing and CNN for effective linking and the author linking module uses graph attention networks to learn from the subgraph of linked venues and papers.

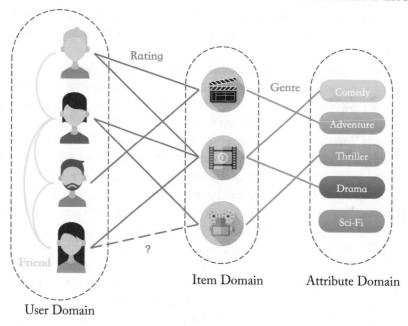

Figure 12.4: Users, items, and attributes are nodes on the graph and interactions between them are edges. In this way, we can convert rating prediction task to link prediction task.

12.4   RECOMMENDER SYSTEMS

In order to take the impact of content information and user-item interaction on recommendation into account at the same time, recommendation systems based on user-item rating graph have received more and more attention. Specifically, this kind of approaches views the users, items, and attributes as nodes of the graph, and the relationships and behavior between users, items, and attributes as edges, and the values on the edges represent the results of the interaction. In this way, as shown in Figure 12.4, the recommendation problem is transformed into a link prediction problem on the graph. Due to the strong representation ability and high interpretability of GNN, GNN-based recommendation methods has been popular.

12.4.1   MATRIX COMPLETION

sRMGCNN (separable Recurrent Muti-Graph CNN) [Monti et al., 2017] was proposed in 2017, which considers the combination of Muti-Graph CNN and RNN. Using the similarity information encoded by the rows and columns of user-item graphs, Muti-Graph CNN could extract local stationary features from the rating matrix. Then these local features are feed into a RNN, the RNN propagates rating values and reconstructs the rating matrix. sRMGCNN inherits the traditional graph convolution methods. It converts a graph to the spectral domain by graph Fourier transformation to guarantee the correctness and convergence. Also, sRMGCNN incorporates matrix factorization model, improves efficiency by factorizing rating matrix into two low-rank matrices.

Different from the spectral convolutional methods, GCMC [van den Berg et al., 2017] is based on spatial GNN which directly aggregates and updates in spatial domain. GCMC can be interpreted as an encoder-decoder model, which gets the user and item nodes embeddings by a graph encoder and obtains the prediction scores by a decoder.

To predict the "instance of" relation between items and collections in web-scale scenarios, PinSage [Ying et al., 2018a] gives a highly efficient model which adopts several useful techniques. The overall structure of PinSage is the same as GraphSage [Hamilton et al., 2017b], it also adopts a sampling strategy to construct computational graphs dynamically. However, in GraphSage, the sampling strategy is random sampling, which is suboptimal when there are large amount of neighbors. PinSage utilizes random walk to generate samples. This technique starts short random walks from target node and assigns weights for the visited nodes. In addition, in order to further improve the computational efficiency of the graph neural network, PinSage designs a computational pipeline to reduce repeated computation. This pipeline method uses the bipartite graph feature of the rating graph to alternately update the representation vector of the items and the representation vector of the collections. Only one half of the node representation is needed for each update.

12.4.2   SOCIAL RECOMMENDATION

Compared to the traditional recommendation setting, social recommendation uses useful social information from social networks of users to enhance the performance. When purchasing online, people are easily affected by others, especially friends. So it is important for recommendation systems to model users' social influence and social correlation. There have been some works using GNN to capture social information.

Wu et al. [2019a] design a neural diffusion model to simulate how the recursive social diffusion process influences users. User embeddings are propagated in social network by a GNN, and combined with pooled item embeddings as output.

To coherently model social network and user-item interaction graph, Fan et al. [2019] propose GraphRec. In this approach, user embeddings aggregate from both social neighborhood and item neighborhood. GraphRec also adopts attention mechanism as aggregator to assign different weights for each node.

Wu et al. [2019b] argue that social effects should not be modeled as static effects, however, they propose to detect four different social effects in recommender situation, including user/item homophily and influence effect. The two kinds of effects jointly affect user preference and item attributes. Wu et al. [2019b] further leverage four GATs to model the four social effects independently.

CHAPTER 13

Applications – Non-Structural Scenarios

In this chapter we will talk about applications on non-structural scenarios such as image, text, programming source code [Allamanis et al., 2018, Li et al., 2016], and multi-agent systems [Hoshen, 2017, Kipf et al., 2018, Sukhbaatar et al., 2016]. We will only give detailed introduction to the first two scenarios due to the length limit. Roughly, there are two ways to apply the graph neural networks on non-structural scenarios: (1) incorporate structural information from other domains to improve the performance, for example using information from knowledge graphs to alleviate the zero-shot problems in image tasks; and (2) infer or assume the relational structure in the scenario and then apply GNN model to solve the problems defined on graphs, such as the method in Zhang et al. [2018c] which models text into graphs.

13.1 IMAGE

13.1.1 IMAGE CLASSIFICATION

Image classification is a very basic and important task in the field of computer vision, which attracts much attention and has many famous datasets like ImageNet [Russakovsky et al., 2015]. Recent progress in image classification benefits from big data and the strong power of GPU computation, which allows us to train a classifier without extracting structural information from images. However, **zero-shot and few-shot learning** are becoming more and more popular in the field of image classification, because most models can achieve similar performance with enough data. There are several works leveraging graph neural networks to incorporate structural information in image classification.

First, knowledge graphs can be used as extra information to guide zero-short recognition classification [Kampffmeyer et al., 2019, Wang et al., 2018c]. Wang et al. [2018c] builds a knowledge graph where each node corresponds to an object category and takes the word embeddings of nodes as input for predicting the classifier of different categories. As over-smoothing effect happens with the deep depth of convolution architecture, the six-layer GCN used in Wang et al. [2018c] would wash out much useful information in the representation. To solve the smoothing problem in the propagation of GCN, Kampffmeyer et al. [2019] managed to use single layer GCN with a larger neighborhood which includes both one-hop and multi-hops nodes in the graph. And it is proved effective in building a zero-shot classifier with existing ones.

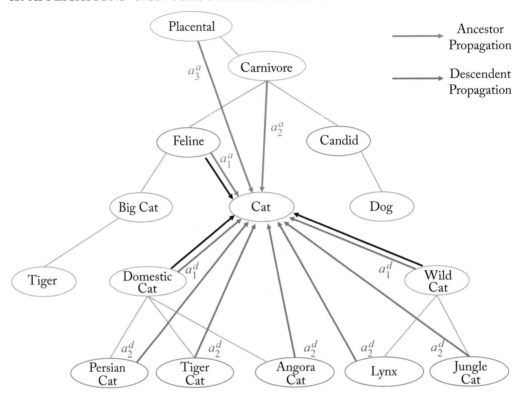

Figure 13.1: The black lines represent the propagation step from previous methods. The red and blues lines represent the propagation step in Kampffmeyer et al. [2019], where the node could aggregate information from ancestor and descendent nodes.

Figure 13.1 shows an example of the propagation step in Kampffmeyer et al. [2019] and Wang et al. [2018c].

Besides the knowledge graph, the similarity between images in the dataset is also helpful for the few-shot learning [Garcia and Bruna, 2018]. Garcia and Bruna [2018] propose to build a weighted fully-connected image network based on the similarity and do message passing in the graph for few-shot recognition.

As most knowledge graphs are large for reasoning, Marino et al. [2017] select some related entities to build a sub-graph based on the result of object detection and apply GGNN to the extracted graph for prediction. Besides, Lee et al. [2018a] propose to construct a new knowledge graph where the entities are all the categories. And, they defined three types of label relations: super-subordinate, positive correlation, and negative correlation and propagate the confidence of labels in the graph directly.

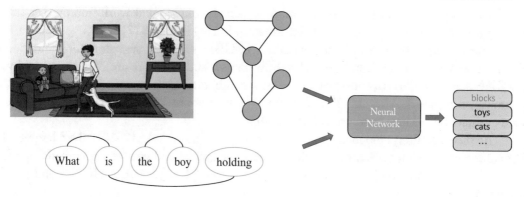

Figure 13.2: The method in Teney et al. [2017] for visual question answering. The scene graph from the picture and the syntactic graph from the question are first constructed and then combined for question answering.

13.1.2   VISUAL REASONING

Computer-vision systems usually need to perform reasoning by incorporating both spatial and semantic information. So it is natural to generate graphs for reasoning tasks.

A typical visual reasoning task is visual question answering (VQA). As shown in Figure 13.2, Teney et al. [2017], respectively, constructs image scene graph and question syntactic graph. Then they apply GGNN to train the embeddings for predicting the final answer. Despite spatial connections among objects, Norcliffebrown et al. [2018] builds the relational graphs conditioned on the questions. With knowledge graphs, Narasimhan et al. [2018] and Wang et al. [2018e] could perform finer relation exploration and more interpretable reasoning process.

Other applications of visual reasoning include object detection, interaction detection, and region classification. In object detection [Gu et al., 2018, Hu et al., 2018], GNNs are used to calculate RoI features. In interaction detection [Jain et al., 2016, Qi et al., 2018], GNNs are used as message passing tools between human and objects. In region classification [Chen et al., 2018c], GNNs perform reasoning on graphs which connects regions and classes.

13.1.3   SEMANTIC SEGMENTATION

Semantic segmentation is a crucial step toward image understanding. The task here is to assign a unique label (or category) to every single pixel in the image, which can be considered as a dense classification problem. However, regions in images are often not grid-like and need nonlocal information, which leads to the failure of traditional CNN. Several works utilized graph-structured data to handle it.

Liang et al. [2016] propose Graph-LSTM to model long-term dependency together with spatial connections by building graphs in form of distance-based superpixel map and applying

LSTM to propagate neighborhood information globally. Subsequent work improved it from the perspective of encoding hierarchical information [Liang et al., 2017].

Furthermore, 3D semantic segmentation (RGBD semantic segmentation) and point clouds classification utilize more geometric information and therefore are hard to model by a 2D CNN. Qi et al. [2017b] construct a K-nearest neighbors (KNN) graph and use a 3D GNN as propagation model. After unrolling for several steps, the prediction model takes the hidden state of each node as input and predict its semantic label.

As there are always too many points, Landrieu and Simonovsky [2018] solved large-scale 3D point clouds segmentation by building superpoint graphs and generating embeddings for them. To classify supernodes, Landrieu and Simonovsky [2018] leverage GGNN and graph convolution.

Wang et al. [2018d] propose to model point interactions through edges. They calculate edge representation vectors by feeding the coordinates of its terminal nodes. Then node embeddings are updated by edge aggregation.

13.2   TEXT

The graph neural networks could be applied to several tasks based on text. It could be applied to both sentence-level tasks (e.g., text classification) as well as word-level tasks (e.g., sequence labeling). We will introduce several major applications on text in the following.

13.2.1   TEXT CLASSIFICATION

Text classification is an important and classical problem in natural language processing. The classical GCN models [Atwood and Towsley, 2016, Defferrard et al., 2016, Hamilton et al., 2017b, Henaff et al., 2015, Kipf and Welling, 2017, Monti et al., 2017] and GAT model [Velickovic et al., 2018] are applied to solve the problem, but they only use the structural information among documents and they don't use much text information.

Peng et al. [2018] propose a graph-CNN-based deep learning model. It first turns texts to graph-of-words, and then conducts the convolution operations in [Niepert et al., 2016] on the word graph.

Zhang et al. [2018c] propose the S-LSTM to encode text. The whole sentence is represented in a single state which contains an overall global state and several sub-states for individual words. It uses the global sentence-level representation for classification tasks.

These methods either view a document or a sentence as a graph of word nodes or rely on the document citation relation to construct the graph. Yao et al. [2019] regard the documents and words as nodes to construct the corpus graph (hence heterogeneous graph) and use the Text GCN to learn embeddings of words and documents.

Sentiment classification could also be regarded as a text classification problem and a Tree-LSTM approach is propose by Tai et al. [2015].

13.2.2   SEQUENCE LABELING

As each node in GNNs has its hidden state, we can utilize the hidden state to address the sequence labeling problem if we consider every word in the sentence as a node. Zhang et al. [2018c] utilize the S-LSTM to label the sequence. They have conducted experiments on POS-tagging and NER tasks and achieves promising performance.

Semantic role labeling is another task of sequence labeling. Marcheggiani and Titov [2017] propose a Syntactic GCN to solve the problem. The Syntactic GCN which operates on the direct graph with labeled edges is a special variant of the GCN [Kipf and Welling, 2017]. It applies edge-wise gates that enable the model to regulate the contribution of each dependency edge. The Syntactic GCNs over syntactic dependency trees are used as sentence encoders to learn latent feature representations of words in the sentence. Marcheggiani and Titov [2017] also reveal that GCNs and LSTMs are functionally complementary in the task. An example of Syntactic GCN could be found in Figure 13.3.

13.2.3   NEURAL MACHINE TRANSLATION

The neural machine translation task is usually considered as a sequence-to-sequence task. Vaswani et al. [2017] introduce the attention mechanisms and replaces the most commonly used recurrent or convolutional layers. In fact, the Transformer assumes a fully connected graph structure between linguistic entities.

One popular application of GNN is to incorporate the syntactic or semantic information into the NMT task. Bastings et al. [2017] utilize the Syntactic GCN on syntax-aware NMT tasks. Marcheggiani et al. [2018] incorporate information about the predicate-argument structure of source sentences (namely, semantic-role representations) using Syntactic GCN and compares the results of incorporating only syntactic or semantic information or both of the information into the task. Beck et al. [2018] utilize the GGNN in syntax-aware NMT. It converts the syntactic dependency graph into a new structure called the Levi graph by turning the edges into additional nodes and thus edge labels can be represented as embeddings.

13.2.4   RELATION EXTRACTION

Extracting semantic relations between entities in texts is an important and well-studied task. Some systems treat this task as a pipeline of two separated tasks, named entity recognition and relation extraction. Miwa and Bansal [2016] propose an end-to-end relation extraction model by using bidirectional sequential and bidirectional tree-structured LSTM-RNNs. Zhang et al. [2018d] propose an extension of graph convolutional networks that is tailored for relation extraction and they have applied a pruning strategy to the input trees.

Zhu et al. [2019a] propose a variant of graph neural network with generated parameters (GP-GNN) for relation extraction. Existing relation extraction methods could easily extract the facts from the text but fail to infer the relations which require multi-hop relational reasoning.

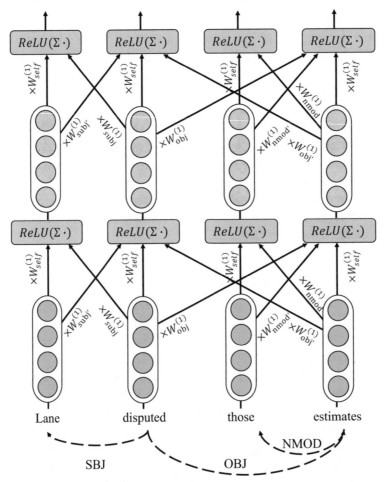

Figure 13.3: An example of Syntactic GCN. This figure shows the example with two Syntactic GCN layers.

GNNs can process multi-hop relational reasoning on graphs but cannot be directly applied on text. So GP-GNN is propose to solve the relational reasoning task on text.

Cross-sentence N-ary relation extraction detects relations among n entities across multiple sentences. Peng et al. [2017] explore a general framework for cross-sentence n-ary relation extraction based on graph LSTMs. It splits the input graph into two DAGs while in this procedure useful information can be lost. Song et al. [2018c] propose a graph-state LSTM model that keeps the original graph structure. Furthermore, the model allows more parallelization to speed up the computation.

13.2.5   EVENT EXTRACTION

Event extraction is an important information extraction task to recognize instances of specified types of events in texts. Nguyen and Grishman [2018] investigate a CNN (which is the Syntactic GCN exactly) based on dependency trees to perform event detection. Liu et al. [2018] propose a Jointly Multiple Events Extraction (JMEE) framework which extracts event triggers and arguments jointly. It uses an attention-based GCN to model graph information and uses shortcut arcs from syntactic structures to enhance information flow.

13.2.6   FACT VERIFICATION

Fact verification (FV) is a challenging task which requires to retrieve relevant evidence from plain text and use the evidence to verify given claims. More specifically, given a claim, an FV system is asked to label it as "SUPPORTED," "REFUTED," or "NOT ENOUGH INFO," which indicate that the evidence support, refute, or is not sufficient for the claim.

Existing FV methods formulate FV as a natural language inference (NLI) [Angeli and Manning, 2014] task. However, they utilize simple evidence combination methods such as concatenating the evidence or just dealing with each evidence-claim pair. These methods are unable to grasp sufficient relational and logical information among the evidence. In fact, many claims require to simultaneously integrate and reason over several pieces of evidence for verification.

To integrate and reason over information from multiple pieces of evidence, [Zhou et al., 2019] propose a graph-based evidence aggregating and reasoning (GEAR) framework. Specifically, it first builds a fully connected evidence graph and encourages information propagation among the evidence. Then, it aggregates the pieces of evidence and adopts a classifier to decide whether the evidence can support, refute, or is not sufficient for the claim.

As shown in Figure 13.4, given a claim and the retrieved evidence, GEAR first utilizes a **sentence encoder** to obtain representations for the claim and the evidence. Then it builds a fully connected evidence graph and proposes an **evidence reasoning network** (ERNet) to propagate information among evidence and reason over the graph. Finally, it utilizes an **evidence aggregator** to infer the final results.

The ERNet used in the evidence reasoning step is a modified version of GAT. More details of ERNet could be found in Zhou et al. [2019].

13.2.7   OTHER APPLICATIONS

GNNs could also be applied to many other applications. There are several works focus on the AMR to text generation task. An S-LSTM based method [Song et al., 2018b] and a GGNN-based method [Beck et al., 2018] have been proposed in this area. Tai et al. [2015] use the Tree-LSTM to model the semantic relatedness of two sentences. And Song et al. [2018a] exploit the Sentence LSTM to solve the multi-hop reading comprehension problem. Another important direction is relational reasoning, relational networks [Santoro et al., 2017], interaction networks [Battaglia et al., 2016], and recurrent relational networks [Palm et al., 2018] are

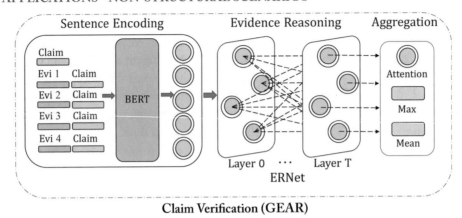

Figure 13.4: The GEAR framework described in Zhou et al. [2019].

proposed to solve the relational reasoning task based on text. The works cited above are not an exhaustive list, and we encourage our readers to find more works and application domains of graph neural networks that they are interested in.

CHAPTER 14

Applications – Other Scenarios

Besides structural and non-structural scenarios, there are some other scenarios where graph neural networks play an important role. In this subsection, we will introduce generative graph models and combinatorial optimization with GNNs.

14.1 GENERATIVE MODELS

Generative models for real-world graphs have drawn significant attention for its important applications including modeling social interactions, discovering new chemical structures, and constructing knowledge graphs. As deep learning methods have powerful ability to learn the implicit distribution of graphs, there is a surge in neural graph generative models recently.

NetGAN [Shchur et al., 2018] is one of the first work to build neural graph generative model, which generates graphs via random walks. It transformed the problem of graph generation to the problem of walk generation which takes the random walks from a specific graph as input and trains a walk generative model using GAN architecture. While the generated graph preserves important topological properties of the original graph, the number of nodes is unable to change in the generating process, which is the same as the original graph. GraphRNN [You et al., 2018b] manages to generate the adjacency matrix of a graph by generating the adjacency vector of each node step by step, which can output required networks with different numbers of nodes.

Instead of generating adjacency matrix sequentially, MolGAN [De Cao and Kipf, 2018] predicts discrete graph structure (the adjacency matrix) at once and utilizes a permutation-invariant discriminator to solve the node variant problem in the adjacency matrix. Besides, it applies a reward network for reinforcement learning-based optimization toward desired chemical properties.

Ma et al. [2018] propose constrained variational autoencoders to ensure the semantic validity of generated graphs. The authors apply penalty terms to regularize the distributions of the existence and types of nodes and edges simultaneously. The regularization focuses on ghost nodes and valence, connectivity and node compatibility.

GCPN [You et al., 2018a] incorporated domain-specific rules through reinforcement learning. To successively construct a molecule graph, GCPN follows current policy to decide whether adding an atom or substructure to an existing molecular graph, or adding a bond to connect exiting atoms. The model is trained by molecular property reward and adversarial loss collectively.

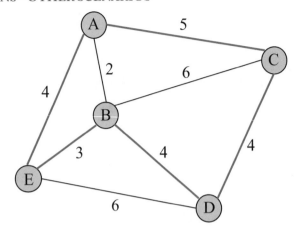

Figure 14.1: A small example of traveling salesman problem (TSP). The nodes denote different cities and edges denote paths between cities. The edge weights are path lengths. The red line shows the shortest possible loop that connects every city.

Li et al. [2018c] propose a model which generates edges and nodes sequentially and utilize a graph neural network to extract the hidden state of the current graph which is used to decide the action in the next step during the sequential generative process.

Rather than small graphs like molecules, Graphite [Grover et al., 2019] is particularly suited for large graphs. The model learns a parameterized distribution of adjacent matrix. Graphite adopts an encoder-decoder architecture, where the encoder is a GNN. For the proposed decoder, the model constructs an intermediate graph and iteratively refine the graph by message passing.

Source code generation is an interesting structured prediction task which requires satisfying semantic and syntactic constraints simultaneously. Brockschmidt et al. [2019] propose to solve this problem by graph generation. They design a novel model which builds a graph from a partial AST by adding edges encoding attribute relationships. A graph neural network performs message passing on the graph helps better guide the generation procedure.

14.2   COMBINATORIAL OPTIMIZATION

Combinatorial optimization problems over graphs are a set of NP-hard problems which attract much attention from scientists of all fields. Some specific problems like traveling salesman problem (TSP) have got various heuristic solutions. Recently, using a DNN for solving such problems has been a hotspot, and some of the solutions further leverage graph neural network because of their graph structure.

Bello et al. [2017] first propose a deep-learning approach to tackle TSP. Their method consists of two parts: a Pointer Network [Vinyals et al., 2015] for parameterizing rewards and a policy gradient [Sutton and Barto, 2018] module for training. This work has been proved to be comparable with traditional approaches. However, Pointer Networks are designed for sequential data like texts, while order-invariant encoders are more appropriate for such work.

Khalil et al. [2017] and Kool et al. [2019] improved the above method by including graph neural networks. The former work first obtain the node embeddings from structure2vec [Dai et al., 2016] then feed them into a Q-learning module for making decisions. The latter one builds an attention-based encoder-decoder system. By replacing reinforcement learning module with an attention-based decoder, it is more efficient for training. These work achieved better performance than previous algorithms, which proved the representation power of graph neural networks. Prates et al. [2019] propose another GNN-based model to tackle TSP. The model assigns a weight for each edge for message passing. The network is trained with a set of dual examples.

Nowak et al. [2018] focus on Quadratic Assignment Problem, i.e., measuring the similarity of two graphs. The GNN-based model learns node embeddings for each graph independently and matches them using attention mechanism. Even in situations where traditional relaxation-based methods appear to suffer, this model offers satisfying performance.

CHAPTER 15

Open Resources

15.1 DATASETS

Many tasks related to graphs are released to test the performance of various graph neural networks. Such tasks are based on the following commonly used datasets.

A series of datasets based on citation networks are as follows:

- **Pubmed** [Yang et al., 2016]

- **Cora** [Yang et al., 2016]

- **Citeseer** [Yang et al., 2016]

- **DBLP** [Tang et al., 2008]

A series of datasets based on Biochemical graphs are as follows:

- **MUTAG** [Debnath et al., 1991]

- **NCI-1** [Wale et al., 2008]

- **PPI** [Zitnik and Leskovec, 2017]

- **D&D** [Dobson and Doig, 2003]

- **PROTEIN** [Borgwardt et al., 2005]

- **PTC** [Toivonen et al., 2003]

A series of datasets based on Social Networks are as follows:

- **Reddit** [Hamilton et al., 2017c]

- **BlogCatalog** [Zafarani and Liu, 2009]

A series of datasets based on Knowledge Graphs are as follows:

- **FB13** [Socher et al., 2013]

- **FB15K** [Bordes et al., 2013]

- **FB15K237** [Toutanova et al., 2015]

- **WN11** [Socher et al., 2013]

- **WN18** [Bordes et al., 2013]

- **WN18RR** [Dettmers et al., 2018]

  A broader range of opensource dataset repositories are as follows:

- **Network Repository**
  A scientific network data repository with interactive visualization and mining tools.
  http://networkrepository.com

- **Graph Kernel Datasets**
  Benchmark datasets for graph kernels.
  https://ls11-www.cs.tu-dortmund.de/staff/morris/graphkerneldatasets

- **Relational Dataset Repository**
  To support the growth of relational machine learning.
  https://relational.fit.cvut.cz

- **Stanford Large Network Dataset Collection**
  The SNAP library is developed to study large social and information networks.
  https://snap.stanford.edu/data/

- **Open Graph Benchmark**
  Open Graph Benchmark (OGB) is a collection of benchmark datasets, data-loaders, and evaluators for graph machine learning in PyTorch.
  https://ogb.stanford.edu/

15.2 IMPLEMENTATIONS

We first list several platforms that provide codes for graph computing in Table 15.1.

Next, we list the hyperlinks of the current opensource implementations of some famous GNN models in Table 15.2.

As the research filed grows rapidly, we recommend our readers the paper list published by our team, GNNPapers (https://github.com/thunlp/gnnpapers), for recent studies.

Table 15.1: Codes for graph computing

Platform	Link	Reference
PyTorch Geometric	https://github.com/rusty1s/pytorch_geometric	[Fey and Lenssen, 2019]
Deep Graph Library	https://github.com/dmlc/dgl	[Wang et al., 2019a]
AliGraph	https://github.com/alibaba/aligraph	[Zhu et al., 2019b]
GraphVite	https://github.com/DeepGraphLearning/graphvite	[Zhu et al., 2019c]
Paddle Graph Learning	https://github.com/PaddlePaddle/PGL	
Euler	https://github.com/alibaba/euler	
Plato	https://github.com/tencent/plato	
CogDL	https://github.com/THUDM/cogdl/	

Table 15.2: Opensource implementations of GNN models

Model	Link
GGNN (2015)	https://github.com/yujiali/ggnn
Neurals FPs (2015)	https://github.com/HIPS/neural-fingerprint
ChebNet (2016)	https://github.com/mdeff/cnn_graph
DNGR (2016)	https://github.com/ShelsonCao/DNGR
SDNE (2016)	https://github.com/suanrong/SDNE
GAE (2016)	https://github.com/limaosen0/Variational-Graph-Auto-Encoders
DRNE (2016)	https://github.com/tadpole/DRNE
Structural RNN (2016)	https://github.com/asheshjain399/RNNexp
DCNN (2016)	https://github.com/jcatw/dcnn
GCN (2017)	https://github.com/tkipf/gcn
CayleyNet (2017)	https://github.com/amoliu/CayleyNet
GraphSage (2017)	https://github.com/williamleif/GraphSAGE
GAT (2017)	https://github.com/PetarV-/GAT
CLN(2017)	https://github.com/trangptm/Column_networks
ECC (2017)	https://github.com/mys007/ecc
MPNNs (2017)	https://github.com/brain-research/mpnn
MoNet (2017)	https://github.com/pierrebaque/GeometricConvolutionsBench
JK-Net (2018)	https://github.com/ShinKyuY/Representation_Learning_on_Graphs_with_Jumping_Knowledge_Networks
SSE (2018)	https://github.com/Hanjun-Dai/steady_state_embedding
LGCN (2018)	https://github.com/divelab/lgcn/
FastGCN (2018)	https://github.com/matenure/FastGCN
DiffPool (2018)	https://github.com/RexYing/diffpool
GraphRNN (2018)	https://github.com/snap-stanford/GraphRNN
MolGAN (2018)	https://github.com/nicola-decao/MolGAN
NetGAN (2018)	https://github.com/danielzuegner/netgan
DCRNN (2018)	https://github.com/liyaguang/DCRNN
ST-GCN (2018)	https://github.com/yysijie/st-gcn
RGCN (2018)	https://github.com/tkipf/relational-gcn
AS-GCN (2018)	https://github.com/huangwb/AS-GCN
DGCN (2018)	https://github.com/ZhuangCY/DGCN
GaAN (2018)	https://github.com/jennyzhang0215/GaAN
DGI (2019)	https://github.com/PetarV-/DGI
GraphWaveNet (2019)	https://github.com/nnzhan/Graph-WaveNet
HAN (2019)	https://github.com/Jhy1993/HAN

CHAPTER 16

Conclusion

Although GNNs have achieved great success in different fields, it is remarkable that GNN models are not good enough to offer satisfying solutions for any graph in any condition. In this section, we will state some open problems for further researches.

Shallow Structure. Traditional DNNs can stack hundreds of layers to get better performance, because deeper structure has more parameters, which improve the expressive power significantly. However, graph neural networks are always shallow, most of which are no more than three layers. As experiments in Li et al. [2018a] show, stacking multiple GCN layers will result in over-smoothing, that is to say, all vertices will converge to the same value. Although some researchers have managed to tackle this problem [Li et al., 2018a, 2016], it remains to be the biggest limitation of GNN. Designing real deep GNN is an exciting challenge for future research, and will be a considerable contribution to the understanding of GNN.

Dynamic Graphs. Another challenging problem is how to deal with graphs with dynamic structures. Static graphs are stable so they can be modeled feasibly, while dynamic graphs introduce changing structures. When edges and nodes appear or disappear, GNN cannot change adaptively. Dynamic GNN is being actively researched on and we believe it to be a big milestone about the stability and adaptability of general GNN.

Non-Structural Scenarios. Although we have discussed the applications of GNN on non-structural scenarios, we found that there is no optimal methods to generate graphs from raw data. In image domain, some work utilizes CNN to obtain feature maps then upsamples them to form superpixels as nodes [Liang et al., 2016], while other ones directly leverage some object detection algorithms to get object nodes. In the text domain [Chen et al., 2018c], some work employs syntactic trees as syntactic graphs while others adopt fully connected graphs. Therefore, finding the best graph generation approach will offer a wider range of fields where GNN could make a contribution.

Scalability. How to apply embedding methods in web-scale conditions like social networks or recommendation systems has been a fatal problem for almost all graph-embedding algorithms, and GNN is not an exception. Scaling up GNN is difficult because many of the core steps are computational consuming in big data environment. There are several examples about this phenomenon. First, graph data are not regular Euclidean, each node has its own neighborhood structure so batches cannot be applied. Then, calculating graph Laplacian is also unfeasible when there are millions of nodes and edges. Moreover, we need to point out that scaling determines whether an algorithm is able to be applied into practical use. Several works have proposed their solutions to this problem [Ying et al., 2018a] and recent research is paying more attention to this direction.

In conclusion, graph neural networks have become powerful and practical tools for machine learning tasks in graph domain. This progress owes to advances in expressive power, model flexibility, and training algorithms. In this book, we give a detailed introduction to graph neural networks. For GNN models, we introduce its variants categorized by graph convolutional networks, graph recurrent networks, graph attention networks and graph residual networks. Moreover, we also summarize several general frameworks to uniformly represent different variants. In terms of application taxonomy, we divide the GNN applications into structural scenarios, non-structural scenarios, and other scenarios, then give a detailed review for applications in each scenario. Finally, we suggest four open problems indicating the major challenges and future research directions of graph neural networks, including model depth, scalability, the ability to deal with dynamic graphs, and non-structural scenarios.

Bibliography

F. Alet, A. K. Jeewajee, M. Bauza, A. Rodriguez, T. Lozano-Perez, and L. P. Kaelbling. 2019. Graph element networks: Adaptive, structured computation and memory. In *Proc. of ICML*. 68

M. Allamanis, M. Brockschmidt, and M. Khademi. 2018. Learning to represent programs with graphs. In *Proc. of ICLR*. 75

G. Angeli and C. D. Manning. 2014. Naturalli: Natural logic inference for common sense reasoning. In *Proc. of EMNLP*, pages 534–545. DOI: 10.3115/v1/d14-1059 81

J. Atwood and D. Towsley. 2016. Diffusion-convolutional neural networks. In *Proc. of NIPS*, pages 1993–2001. 2, 26, 30, 78

D. Bahdanau, K. Cho, and Y. Bengio. 2015. Neural machine translation by jointly learning to align and translate. In *Proc. of ICLR*. 39

J. Bastings, I. Titov, W. Aziz, D. Marcheggiani, and K. Simaan. 2017. Graph convolutional encoders for syntax-aware neural machine translation. In *Proc. of EMNLP*, pages 1957–1967. DOI: 10.18653/v1/d17-1209 79

P. Battaglia, R. Pascanu, M. Lai, D. J. Rezende, et al. 2016. Interaction networks for learning about objects, relations and physics. In *Proc. of NIPS*, pages 4502–4510. 1, 59, 63, 67, 81

P. W. Battaglia, J. B. Hamrick, V. Bapst, A. Sanchez-Gonzalez, V. Zambaldi, M. Malinowski, A. Tacchetti, D. Raposo, A. Santoro, R. Faulkner, et al. 2018. Relational inductive biases, deep learning, and graph networks. *ArXiv Preprint ArXiv:1806.01261*. 3, 59, 62, 63, 64

D. Beck, G. Haffari, and T. Cohn. 2018. Graph-to-sequence learning using gated graph neural networks. In *Proc. of ACL*, pages 273–283. DOI: 10.18653/v1/p18-1026 49, 79, 81

I. Bello, H. Pham, Q. V. Le, M. Norouzi, and S. Bengio. 2017. Neural combinatorial optimization with reinforcement learning. In *Proc. of ICLR*. 84

Y. Bengio, P. Simard, P. Frasconi, et al. 1994. Learning long-term dependencies with gradient descent is difficult. *IEEE TNN*, 5(2):157–166. DOI: 10.1109/72.279181 17

M. Berlingerio, M. Coscia, and F. Giannotti. 2011. Finding redundant and complementary communities in multidimensional networks. In *Proc. of CIKM*, pages 2181–2184. ACM. DOI: 10.1145/2063576.2063921 52

A. Bordes, N. Usunier, A. Garcia-Duran, J. Weston, and O. Yakhnenko. 2013. Translating embeddings for modeling multi-relational data. In *Proc. of NIPS*, pages 2787–2795. 87, 88

K. M. Borgwardt, C. S. Ong, S. Schönauer, S. Vishwanathan, A. J. Smola, and H.-P. Kriegel. 2005. Protein function prediction via graph kernels. *Bioinformatics*, 21(suppl_1):i47–i56. DOI: 10.1093/bioinformatics/bti1007 87

D. Boscaini, J. Masci, E. Rodolà, and M. Bronstein. Learning shape correspondence with anisotropic convolutional neural networks. In *Proc. of NIPS*, pages 3189–3197. 2, 30

J. Bradshaw, M. J. Kusner, B. Paige, M. H. Segler, and J. M. Hernández-Lobato. 2019. A generative model for electron paths. In *Proc. of ICLR*. 70

M. Brockschmidt, M. Allamanis, A. L. Gaunt, and O. Polozov. 2019. Generative code modeling with graphs. In *Proc. of ICLR*. 84

M. M. Bronstein, J. Bruna, Y. LeCun, A. Szlam, and P. Vandergheynst. 2017. Geometric deep learning: going beyond euclidean data. *IEEE SPM*, 34(4):18–42. DOI: 10.1109/msp.2017.2693418 2

J. Bruna, W. Zaremba, A. Szlam, and Y. Lecun. 2014. Spectral networks and locally connected networks on graphs. In *Proc. of ICLR*. 23, 59

A. Buades, B. Coll, and J.-M. Morel. 2005. A non-local algorithm for image denoising. In *Proc. of CVPR*, 2:60–65. IEEE. DOI: 10.1109/cvpr.2005.38 60, 61

H. Cai, V. W. Zheng, and K. C.-C. Chang. 2018. A comprehensive survey of graph embedding: Problems, techniques, and applications. *IEEE TKDE*, 30(9):1616–1637. DOI: 10.1109/tkde.2018.2807452 2

S. Cao, W. Lu, and Q. Xu. 2016. Deep neural networks for learning graph representations. In *Proc. of AAAI*. 56

M. Chang, T. Ullman, A. Torralba, and J. B. Tenenbaum. 2017. A compositional object-based approach to learning physical dynamics. In *Proc. of ICLR*. 59, 63

J. Chen, T. Ma, and C. Xiao. 2018a. FastGCN: Fast learning with graph convolutional networks via importance sampling. In *Proc. of ICLR*. 54

J. Chen, J. Zhu, and L. Song. 2018b. Stochastic training of graph convolutional networks with variance reduction. In *Proc. of ICML*, pages 941–949. 55

X. Chen, L.-J. Li, L. Fei-Fei, and A. Gupta. 2018c. Iterative visual reasoning beyond convolutions. In *Proc. of CVPR*, pages 7239–7248. DOI: 10.1109/cvpr.2018.00756 77, 91

X. Chen, G. Yu, J. Wang, C. Domeniconi, Z. Li, and X. Zhang. 2019. Activehne: Active heterogeneous network embedding. In *Proc. of IJCAI*. DOI: 10.24963/ijcai.2019/294 48

J. Cheng, L. Dong, and M. Lapata. 2016. Long short-term memory-networks for machine reading. In *Proc. of EMNLP*, pages 551–561. DOI: 10.18653/v1/d16-1053 39

K. Cho, B. Van Merrienboer, C. Gulcehre, D. Bahdanau, F. Bougares, H. Schwenk, and Y. Bengio. 2014. Learning phrase representations using RNN encoder—decoder for statistical machine translation. In *Proc. of EMNLP*, pages 1724–1734. DOI: 10.3115/v1/d14-1179 17, 33, 60

F. R. Chung and F. C. Graham. 1997. *Spectral Graph Theory*. American Mathematical Society. DOI: 10.1090/cbms/092 1

P. Cui, X. Wang, J. Pei, and W. Zhu. 2018. A survey on network embedding. *IEEE TKDE*. DOI: 10.1109/TKDE.2018.2849727 2

H. Dai, B. Dai, and L. Song. 2016. Discriminative embeddings of latent variable models for structured data. In *Proc. of ICML*, pages 2702–2711. 59, 85

H. Dai, Z. Kozareva, B. Dai, A. Smola, and L. Song. 2018. Learning steady-states of iterative algorithms over graphs. In *Proc. of ICML*, pages 1114–1122. 55

N. De Cao and T. Kipf. 2018. MolGAN: An implicit generative model for small molecular graphs. *ICML Workshop on Theoretical Foundations and Applications of Deep Generative Models*. 83

A. K. Debnath, R. L. Lopez de Compadre, G. Debnath, A. J. Shusterman, and C. Hansch. 1991. Structure-activity relationship of mutagenic aromatic and heteroaromatic nitro compounds. Correlation with molecular orbital energies and hydrophobicity. *Journal of Medicinal Chemistry*, 34(2):786–797. DOI: 10.1021/jm00106a046 87

M. Defferrard, X. Bresson, and P. Vandergheynst. 2016. Convolutional neural networks on graphs with fast localized spectral filtering. In *Proc. of NIPS*, pages 3844–3852. 24, 59, 78

T. Dettmers, P. Minervini, P. Stenetorp, and S. Riedel. 2018. Convolutional 2D knowledge graph embeddings. In *Proc. of AAAI*. 71, 88

K. Do, T. Tran, and S. Venkatesh. 2019. Graph transformation policy network for chemical reaction prediction. In *Proc. of SIGKDD*, pages 750–760. ACM. DOI: 10.1145/3292500.3330958 70

P. D. Dobson and A. J. Doig. 2003. Distinguishing enzyme structures from non-enzymes without alignments. *Journal of Molecular Biology*, 330(4):771–783. DOI: 10.1016/s0022-2836(03)00628-4 87

D. K. Duvenaud, D. Maclaurin, J. Aguileraiparraguirre, R. Gomezbombarelli, T. D. Hirzel, A. Aspuruguzik, and R. P. Adams. 2015. Convolutional networks on graphs for learning molecular fingerprints. In *Proc. of NIPS*, pages 2224–2232. 25, 59, 68

W. Fan, Y. Ma, Q. Li, Y. He, E. Zhao, J. Tang, and D. Yin. 2019. Graph neural networks for social recommendation. In *Proc. of WWW*, pages 417–426. ACM. DOI: 10.1145/3308558.3313488 74

M. Fey and J. E. Lenssen. 2019. Fast graph representation learning with PyTorch Geometric. In *ICLR Workshop on Representation Learning on Graphs and Manifolds*. 89

A. Fout, J. Byrd, B. Shariat, and A. Ben-Hur. 2017. Protein interface prediction using graph convolutional networks. In *Proc. of NIPS*, pages 6530–6539. 1, 70

H. Gao, Z. Wang, and S. Ji. 2018. Large-scale learnable graph convolutional networks. In *Proc. of SIGKDD*, pages 1416–1424. ACM. DOI: 10.1145/3219819.3219947 29

V. Garcia and J. Bruna. 2018. Few-shot learning with graph neural networks. In *Proc. of ICLR*. 76

J. Gehring, M. Auli, D. Grangier, and Y. N. Dauphin. 2017. A convolutional encoder model for neural machine translation. In *Proc. of ACL*, 1:123–135. DOI: 10.18653/v1/p17-1012 39

J. Gilmer, S. S. Schoenholz, P. F. Riley, O. Vinyals, and G. E. Dahl. 2017. Neural message passing for quantum chemistry. In *Proc. of ICML*, pages 1263–1272. 3, 59, 60, 62, 63, 64

M. Gori, G. Monfardini, and F. Scarselli. 2005. A new model for learning in graph domains. In *Proc. of IJCNN*, pages 729–734. DOI: 10.1109/ijcnn.2005.1555942 19

P. Goyal and E. Ferrara. 2018. Graph embedding techniques, applications, and performance: A survey. *Knowledge-Based Systems*, 151:78–94. DOI: 10.1016/j.knosys.2018.03.022 2

J. L. Gross and J. Yellen. 2004. *Handbook of Graph Theory*. CRC Press. DOI: 10.1201/9780203490204 49

A. Grover and J. Leskovec. 2016. node2vec: Scalable feature learning for networks. In *Proc. of SIGKDD*, pages 855–864. ACM. DOI: 10.1145/2939672.2939754 2

A. Grover, A. Zweig, and S. Ermon. 2019. Graphite: Iterative generative modeling of graphs. In *Proc. of ICML*. 84

J. Gu, H. Hu, L. Wang, Y. Wei, and J. Dai. 2018. Learning region features for object detection. In *Proc. of ECCV*, pages 381–395. DOI: 10.1007/978-3-030-01258-8_24 77

T. Hamaguchi, H. Oiwa, M. Shimbo, and Y. Matsumoto. 2017. Knowledge transfer for out-of-knowledge-base entities: A graph neural network approach. In *Proc. of IJCAI*, pages 1802–1808. DOI: 10.24963/ijcai.2017/250 1, 72

W. L. Hamilton, R. Ying, and J. Leskovec. 2017a. Representation learning on graphs: Methods and applications. *IEEE Data(base) Engineering Bulletin*, 40:52–74. 2

W. L. Hamilton, Z. Ying, and J. Leskovec. 2017b. Inductive representation learning on large graphs. In *Proc. of NIPS*, pages 1024–1034. 1, 31, 32, 53, 67, 74, 78

W. L. Hamilton, J. Zhang, C. Danescu-Niculescu-Mizil, D. Jurafsky, and J. Leskovec. 2017c. Loyalty in online communities. In *Proc. of ICWSM*. 87

D. K. Hammond, P. Vandergheynst, and R. Gribonval. 2011. Wavelets on graphs via spectral graph theory. *Applied and Computational Harmonic Analysis*, 30(2):129–150. DOI: 10.1016/j.acha.2010.04.005 24

J. B. Hamrick, K. Allen, V. Bapst, T. Zhu, K. R. Mckee, J. B. Tenenbaum, and P. Battaglia. 2018. Relational inductive bias for physical construction in humans and machines. *Cognitive Science*. 63

K. He, X. Zhang, S. Ren, and J. Sun. 2016a. Deep residual learning for image recognition. In *Proc. of CVPR*, pages 770–778. DOI: 10.1109/cvpr.2016.90 43, 61

K. He, X. Zhang, S. Ren, and J. Sun. 2016b. Identity mappings in deep residual networks. In *Proc. of ECCV*, pages 630–645. Springer. DOI: 10.1007/978-3-319-46493-0_38 31, 45

M. Henaff, J. Bruna, and Y. Lecun. 2015. Deep convolutional networks on graph-structured data. *ArXiv: Preprint, ArXiv:1506.05163*. 23, 78

S. Hochreiter and J. Schmidhuber. 1997. Long short-term memory. *Neural Computation*, 9(8):1735–1780. DOI: 10.1162/neco.1997.9.8.1735 17, 33

S. Hochreiter, Y. Bengio, P. Frasconi, J. Schmidhuber, et al., 2001. Gradient flow in recurrent nets: The difficulty of learning long-term dependencies. *A Field Guide to Dynamical Recurrent Neural Networks*. IEEE Press. 17

Y. Hoshen. 2017. Vain: Attentional multi-agent predictive modeling. In *Proc. of NIPS*, pages 2701–2711. 59, 60, 67, 75

H. Hu, J. Gu, Z. Zhang, J. Dai, and Y. Wei. 2018. Relation networks for object detection. In *Proc. of CVPR*, pages 3588–3597. DOI: 10.1109/cvpr.2018.00378 77

G. Huang, Z. Liu, L. Van Der Maaten, and K. Q. Weinberger. 2017. Densely connected convolutional networks. In *Proc. of CVPR*, pages 4700–4708. DOI: 10.1109/cvpr.2017.243 45

W. Huang, T. Zhang, Y. Rong, and J. Huang. 2018. Adaptive sampling towards fast graph representation learning. In *Proc. of NeurIPS*, pages 4563–4572. 54

T. J. Hughes. 2012. *The Finite Element Method: Linear Static and Dynamic Finite Element Analysis*. Courier Corporation. 68

A. Jain, A. R. Zamir, S. Savarese, and A. Saxena. 2016. Structural-RNN: Deep learning on spatio-temporal graphs. In *Proc. of CVPR*, pages 5308–5317. DOI: 10.1109/cvpr.2016.573 51, 77

W. Jin, R. Barzilay, and T. Jaakkola. 2018. Junction tree variational autoencoder for molecular graph generation. In *Proc. of ICML*. 69

W. Jin, K. Yang, R. Barzilay, and T. Jaakkola. 2019. Learning multimodal graph-to-graph translation for molecular optimization. In *Proc. of ICLR*. 69

M. Kampffmeyer, Y. Chen, X. Liang, H. Wang, Y. Zhang, and E. P. Xing. 2019. Rethinking knowledge graph propagation for zero-shot learning. In *Proc. of CVPR*. DOI: 10.1109/cvpr.2019.01175 47, 75, 76

S. Kearnes, K. McCloskey, M. Berndl, V. Pande, and P. Riley. 2016. Molecular graph convolutions: Moving beyond fingerprints. *Journal of Computer-Aided Molecular Design*, 30(8):595–608. DOI: 10.1007/s10822-016-9938-8 59, 69

E. Khalil, H. Dai, Y. Zhang, B. Dilkina, and L. Song. 2017. Learning combinatorial optimization algorithms over graphs. In *Proc. of NIPS*, pages 6348–6358. 1, 59, 85

M. A. Khamsi and W. A. Kirk. 2011. *An Introduction to Metric Spaces and Fixed Point Theory*, volume 53. John Wiley & Sons. DOI: 10.1002/9781118033074 20

M. R. Khan and J. E. Blumenstock. 2019. Multi-GCN: Graph convolutional networks for multi-view networks, with applications to global poverty. *ArXiv Preprint ArXiv:1901.11213*. DOI: 10.1609/aaai.v33i01.3301606 52

T. Kipf, E. Fetaya, K. Wang, M. Welling, and R. S. Zemel. 2018. Neural relational inference for interacting systems. In *Proc. of ICML*, pages 2688–2697. 63, 67, 75

T. N. Kipf and M. Welling. 2016. Variational graph auto-encoders. In *Proc. of NIPS*. 56

T. N. Kipf and M. Welling. 2017. Semi-supervised classification with graph convolutional networks. In *Proc. of ICLR*. 1, 2, 24, 30, 31, 43, 48, 53, 59, 67, 78, 79

W. Kool, H. van Hoof, and M. Welling. 2019. Attention, learn to solve routing problems! In *Proc. of ICLR*. https://openreview.net/forum?id=ByxBFsRqYm 85

A. Krizhevsky, I. Sutskever, and G. E. Hinton. 2012. Imagenet classification with deep convolutional neural networks. In *Proc. of NIPS*, pages 1097–1105. DOI: 10.1145/3065386 17

L. Landrieu and M. Simonovsky. 2018. Large-scale point cloud semantic segmentation with superpoint graphs. In *Proc. of CVPR*, pages 4558–4567. DOI: 10.1109/cvpr.2018.00479 78

Y. LeCun, L. Bottou, Y. Bengio, and P. Haffner. 1998. Gradient-based learning applied to document recognition. *Proc. of the IEEE*, 86(11):2278–2324. DOI: 10.1109/5.726791 1, 17

Y. LeCun, Y. Bengio, and G. Hinton. 2015. Deep learning. *Nature*, 521(7553):436. DOI: 10.1038/nature14539 1

C. Lee, W. Fang, C. Yeh, and Y. F. Wang. 2018a. Multi-label zero-shot learning with structured knowledge graphs. In *Proc. of CVPR*, pages 1576–1585. DOI: 10.1109/cvpr.2018.00170 76

G.-H. Lee, W. Jin, D. Alvarez-Melis, and T. S. Jaakkola. 2019. Functional transparency for structured data: A game-theoretic approach. In *Proc. of ICML*. 69

J. B. Lee, R. A. Rossi, S. Kim, N. K. Ahmed, and E. Koh. 2018b. Attention models in graphs: A survey. *ArXiv Preprint ArXiv:1807.07984*. DOI: 10.1145/3363574 3

F. W. Levi. 1942. *Finite Geometrical Systems: Six Public Lectures Delivered in February, 1940, at the University of Calcutta*. The University of Calcutta. 49

G. Li, M. Muller, A. Thabet, and B. Ghanem. 2019. DeepGCNs: Can GCNs go as deep as CNNs? In *Proc. of ICCV*. 45, 46

Q. Li, Z. Han, and X.-M. Wu. 2018a. Deeper insights into graph convolutional networks for semi-supervised learning. In *Proc. of AAAI*. 55, 91

R. Li, S. Wang, F. Zhu, and J. Huang. 2018b. Adaptive graph convolutional neural networks. In *Proc. of AAAI*. 25

Y. Li, D. Tarlow, M. Brockschmidt, and R. S. Zemel. 2016. Gated graph sequence neural networks. In *Proc. of ICLR*. 22, 33, 59, 75, 91

Y. Li, O. Vinyals, C. Dyer, R. Pascanu, and P. Battaglia. 2018c. Learning deep generative models of graphs. In *Proc. of ICLR Workshop*. 83

Y. Li, R. Yu, C. Shahabi, and Y. Liu. 2018d. Diffusion convolutional recurrent neural network: Data-driven traffic forecasting. In *Proc. of ICLR*. DOI: 10.1109/trustcom/bigdatase.2019.00096 50

X. Liang, X. Shen, J. Feng, L. Lin, and S. Yan. 2016. Semantic object parsing with graph LSTM. In *Proc. of ECCV*, pages 125–143. DOI: 10.1007/978-3-319-46448-0_8 36, 77, 91

X. Liang, L. Lin, X. Shen, J. Feng, S. Yan, and E. P. Xing. 2017. Interpretable structure-evolving LSTM. In *Proc. of CVPR*, pages 2175–2184. DOI: 10.1109/cvpr.2017.234 78

X. Liu, Z. Luo, and H. Huang. 2018. Jointly multiple events extraction via attention-based graph information aggregation. In *Proc. of EMNLP*. DOI: 10.18653/v1/d18-1156 81

T. Ma, J. Chen, and C. Xiao. 2018. Constrained generation of semantically valid graphs via regularizing variational autoencoders. In *Proc. of NeurIPS*, pages 7113–7124. 83

Y. Ma, S. Wang, C. C. Aggarwal, D. Yin, and J. Tang. 2019. Multi-dimensional graph convolutional networks. In *Proc. of SDM*, pages 657–665. DOI: 10.1137/1.9781611975673.74 52

D. Marcheggiani and I. Titov. 2017. Encoding sentences with graph convolutional networks for semantic role labeling. In *Proc. of EMNLP*, pages 1506–1515. DOI: 10.18653/v1/d17-1159 79

D. Marcheggiani, J. Bastings, and I. Titov. 2018. Exploiting semantics in neural machine translation with graph convolutional networks. In *Proc. of NAACL*. DOI: 10.18653/v1/n18-2078 79

K. Marino, R. Salakhutdinov, and A. Gupta. 2017. The more you know: Using knowledge graphs for image classification. In *Proc. of CVPR*, pages 20–28. DOI: 10.1109/cvpr.2017.10 76

J. Masci, D. Boscaini, M. Bronstein, and P. Vandergheynst. 2015. Geodesic convolutional neural networks on Riemannian manifolds. In *Proc. of ICCV Workshops*, pages 37–45. DOI: 10.1109/iccvw.2015.112 2, 30

T. Mikolov, K. Chen, G. Corrado, and J. Dean. 2013. Efficient estimation of word representations in vector space. In *Proc. of ICLR*. 2

M. Miwa and M. Bansal. 2016. End-to-end relation extraction using LSTMs on sequences and tree structures. In *Proc. of ACL*, pages 1105–1116. DOI: 10.18653/v1/p16-1105 79

F. Monti, D. Boscaini, J. Masci, E. Rodola, J. Svoboda, and M. M. Bronstein. 2017. Geometric deep learning on graphs and manifolds using mixture model CNNs. In *Proc. of CVPR*, pages 5425–5434. DOI: 10.1109/cvpr.2017.576 2, 30, 73, 78

M. Narasimhan, S. Lazebnik, and A. G. Schwing. 2018. Out of the box: Reasoning with graph convolution nets for factual visual question answering. In *Proc. of NeurIPS*, pages 2654–2665. 77

D. Nathani, J. Chauhan, C. Sharma, and M. Kaul. 2019. Learning attention-based embeddings for relation prediction in knowledge graphs. In *Proc. of ACL*. DOI: 10.18653/v1/p19-1466 72

T. H. Nguyen and R. Grishman. 2018. Graph convolutional networks with argument-aware pooling for event detection. In *Proc. of AAAI*. 81

M. Niepert, M. Ahmed, and K. Kutzkov. 2016. Learning convolutional neural networks for graphs. In *Proc. of ICML*, pages 2014–2023. 26, 78

W. Norcliffebrown, S. Vafeias, and S. Parisot. 2018. Learning conditioned graph structures for interpretable visual question answering. In *Proc. of NeurIPS*, pages 8334–8343. 77

A. Nowak, S. Villar, A. S. Bandeira, and J. Bruna. 2018. Revised note on learning quadratic assignment with graph neural networks. In *Proc. of IEEE DSW*, pages 1–5. IEEE. DOI: 10.1109/dsw.2018.8439919 85

R. Palm, U. Paquet, and O. Winther. 2018. Recurrent relational networks. In *Proc. of NeurIPS*, pages 3368–3378. 81

S. Pan, R. Hu, G. Long, J. Jiang, L. Yao, and C. Zhang. 2018. Adversarially regularized graph autoencoder for graph embedding. In *Proc. of IJCAI*. DOI: 10.24963/ijcai.2018/362 56

E. E. Papalexakis, L. Akoglu, and D. Ience. 2013. Do more views of a graph help? Community detection and clustering in multi-graphs. In *Proc. of FUSION*, pages 899–905. IEEE. 52

H. Peng, J. Li, Y. He, Y. Liu, M. Bao, L. Wang, Y. Song, and Q. Yang. 2018. Large-scale hierarchical text classification with recursively regularized deep graph-CNN. In *Proc. of WWW*, pages 1063–1072. DOI: 10.1145/3178876.3186005 78

H. Peng, J. Li, Q. Gong, Y. Song, Y. Ning, K. Lai, and P. S. Yu. 2019. Fine-grained event categorization with heterogeneous graph convolutional networks. In *Proc. of IJCAI*. DOI: 10.24963/ijcai.2019/449 48

N. Peng, H. Poon, C. Quirk, K. Toutanova, and W.-t. Yih. 2017. Cross-sentence N-ary relation extraction with graph LSTMs. *TACL*, 5:101–115. DOI: 10.1162/tacl_a_00049 35, 80

B. Perozzi, R. Al-Rfou, and S. Skiena. 2014. Deepwalk: Online learning of social representations. In *Proc. of SIGKDD*, pages 701–710. ACM. DOI: 10.1145/2623330.2623732 2

T. Pham, T. Tran, D. Phung, and S. Venkatesh. 2017. Column networks for collective classification. In *Proc. of AAAI*. 43

M. Prates, P. H. Avelar, H. Lemos, L. C. Lamb, and M. Y. Vardi. 2019. Learning to solve NP-complete problems: A graph neural network for decision TSP. In *Proc. of AAAI*, 33:4731–4738. DOI: 10.1609/aaai.v33i01.33014731 85

C. R. Qi, H. Su, K. Mo, and L. J. Guibas. 2017a. PointNet: Deep learning on point sets for 3D classification and segmentation. In *Proc. of CVPR*, 1(2):4. DOI: 10.1109/cvpr.2017.16 59

S. Qi, W. Wang, B. Jia, J. Shen, and S.-C. Zhu. 2018. Learning human-object interactions by graph parsing neural networks. In *Proc. of ECCV*, pages 401–417. DOI: 10.1007/978-3-030-01240-3_25 77

X. Qi, R. Liao, J. Jia, S. Fidler, and R. Urtasun. 2017b. 3D graph neural networks for RGBD semantic segmentation. In *Proc. of CVPR*, pages 5199–5208. DOI: 10.1109/iccv.2017.556 78

A. Rahimi, T. Cohn, and T. Baldwin. 2018. Semi-supervised user geolocation via graph convolutional networks. In *Proc. of ACL*, 1:2009–2019. DOI: 10.18653/v1/p18-1187 43, 67

D. Raposo, A. Santoro, D. G. T. Barrett, R. Pascanu, T. P. Lillicrap, and P. Battaglia. 2017. Discovering objects and their relations from entangled scene representations. In *Proc. of ICLR*. 59, 64

S. Rhee, S. Seo, and S. Kim. 2018. Hybrid approach of relation network and localized graph convolutional filtering for breast cancer subtype classification. In *Proc. of IJCAI*. DOI: 10.24963/ijcai.2018/490 70

O. Russakovsky, J. Deng, H. Su, J. Krause, S. Satheesh, S. Ma, Z. Huang, A. Karpathy, A. Khosla, M. Bernstein, et al. 2015. ImageNet large scale visual recognition challenge. In *Proc. of IJCV*, 115(3):211–252. DOI: 10.1007/s11263-015-0816-y 75

A. Sanchez, N. Heess, J. T. Springenberg, J. Merel, R. Hadsell, M. A. Riedmiller, and P. Battaglia. 2018. Graph networks as learnable physics engines for inference and control. In *Proc. of ICLR*, pages 4467–4476. 1, 63, 64, 67

A. Santoro, D. Raposo, D. G. Barrett, M. Malinowski, R. Pascanu, P. Battaglia, and T. Lillicrap. 2017. A simple neural network module for relational reasoning. In *Proc. of NIPS*, pages 4967–4976. 59, 63, 81

F. Scarselli, A. C. Tsoi, M. Gori, and M. Hagenbuchner. 2004. Graphical-based learning environments for pattern recognition. In *Proc. of Joint IAPR International Workshops on SPR and SSPR*, pages 42–56. DOI: 10.1007/978-3-540-27868-9_4 19

F. Scarselli, M. Gori, A. C. Tsoi, M. Hagenbuchner, and G. Monfardini. 2009. The graph neural network model. *IEEE TNN*, pages 61–80. DOI: 10.1109/tnn.2008.2005605 19, 20, 47, 62

M. Schlichtkrull, T. N. Kipf, P. Bloem, R. van den Berg, I. Titov, and M. Welling. 2018. Modeling relational data with graph convolutional networks. In *Proc. of ESWC*, pages 593–607. Springer. DOI: 10.1007/978-3-319-93417-4_38 22, 50, 71

K. T. Schütt, F. Arbabzadah, S. Chmiela, K. R. Müller, and A. Tkatchenko. 2017. Quantum-chemical insights from deep tensor neural networks. *Nature Communications*, 8:13890. DOI: 10.1038/ncomms13890 59

C. Shang, Y. Tang, J. Huang, J. Bi, X. He, and B. Zhou. 2019a. End-to-end structure-aware convolutional networks for knowledge base completion. In *Proc. of AAAI*, 33:3060–3067. DOI: 10.1609/aaai.v33i01.33013060 71

J. Shang, T. Ma, C. Xiao, and J. Sun. 2019b. Pre-training of graph augmented transformers for medication recommendation. In *Proc. of IJCAI*. DOI: 10.24963/ijcai.2019/825 70

J. Shang, C. Xiao, T. Ma, H. Li, and J. Sun. 2019c. GameNet: Graph augmented memory networks for recommending medication combination. In *Proc. of AAAI*, 33:1126–1133. DOI: 10.1609/aaai.v33i01.33011126 70

O. Shchur, D. Zugner, A. Bojchevski, and S. Gunnemann. 2018. NetGAN: Generating graphs via random walks. In *Proc. of ICML*, pages 609–618. 83

M. Simonovsky and N. Komodakis. 2017. Dynamic edge-conditioned filters in convolutional neural networks on graphs. In *Proc. CVPR*, pages 3693–3702. DOI: 10.1109/cvpr.2017.11 55

K. Simonyan and A. Zisserman. 2014. Very deep convolutional networks for large-scale image recognition. *ArXiv Preprint ArXiv:1409.1556*. 17

R. Socher, D. Chen, C. D. Manning, and A. Ng. 2013. Reasoning with neural tensor networks for knowledge base completion. In *Proc. of NIPS*, pages 926–934. 87, 88

L. Song, Z. Wang, M. Yu, Y. Zhang, R. Florian, and D. Gildea. 2018a. Exploring graph-structured passage representation for multi-hop reading comprehension with graph neural networks. *ArXiv Preprint ArXiv:1809.02040*. 81

L. Song, Y. Zhang, Z. Wang, and D. Gildea. 2018b. A graph-to-sequence model for AMR-to-text generation. In *Proc. of ACL*, pages 1616–1626. DOI: 10.18653/v1/p18-1150 81

L. Song, Y. Zhang, Z. Wang, and D. Gildea. 2018c. N-ary relation extraction using graph state LSTM. In *Proc. of EMNLP*, pages 2226–2235. DOI: 10.18653/v1/d18-1246 80

S. Sukhbaatar, R. Fergus, et al. 2016. Learning multiagent communication with backpropagation. In *Proc. of NIPS*, pages 2244–2252. 59, 67, 75

Y. Sun, N. Bui, T.-Y. Hsieh, and V. Honavar. 2018. Multi-view network embedding via graph factorization clustering and co-regularized multi-view agreement. In *IEEE ICDMW*, pages 1006–1013. DOI: 10.1109/icdmw.2018.00145 52

R. S. Sutton and A. G. Barto. 2018. *Reinforcement Learning: An Introduction*. MIT Press. DOI: 10.1109/tnn.1998.712192 85

C. Szegedy, W. Liu, Y. Jia, P. Sermanet, S. Reed, D. Anguelov, D. Erhan, V. Vanhoucke, and A. Rabinovich. 2015. Going deeper with convolutions. In *Proc. of CVPR*, pages 1–9. DOI: 10.1109/cvpr.2015.7298594 17

K. S. Tai, R. Socher, and C. D. Manning. 2015. Improved semantic representations from tree-structured long short-term memory networks. In *Proc. of IJCNLP*, pages 1556–1566. DOI: 10.3115/v1/p15-1150 34, 78, 81

J. Tang, J. Zhang, L. Yao, J. Li, L. Zhang, and Z. Su. 2008. Arnetminer: Extraction and mining of academic social networks. In *Proc. of SIGKDD*, pages 990–998. DOI: 10.1145/1401890.1402008 87

J. Tang, M. Qu, M. Wang, M. Zhang, J. Yan, and Q. Mei. 2015. Line: Large-scale information network embedding. In *Proc. of WWW*, pages 1067–1077. DOI: 10.1145/2736277.2741093 2

D. Teney, L. Liu, and A. V. Den Hengel. 2017. Graph-structured representations for visual question answering. In *Proc. of CVPR*, pages 3233–3241. DOI: 10.1109/cvpr.2017.344 77

H. Toivonen, A. Srinivasan, R. D. King, S. Kramer, and C. Helma. 2003. Statistical evaluation of the predictive toxicology challenge 2000–2001. *Bioinformatics*, 19(10):1183–1193. DOI: 10.1093/bioinformatics/btg130 87

C. Tomasi and R. Manduchi. 1998. Bilateral filtering for gray and color images. In *Computer Vision*, pages 839–846. IEEE. DOI: 10.1109/iccv.1998.710815 61

K. Toutanova, D. Chen, P. Pantel, H. Poon, P. Choudhury, and M. Gamon. 2015. Representing text for joint embedding of text and knowledge bases. In *Proc. of EMNLP*, pages 1499–1509. DOI: 10.18653/v1/d15-1174 87

K. Tu, P. Cui, X. Wang, P. S. Yu, and W. Zhu. 2018. Deep recursive network embedding with regular equivalence. In *Proc. of SIGKDD*. DOI: 10.1145/3219819.3220068 56

R. van den Berg, T. N. Kipf, and M. Welling. 2017. Graph convolutional matrix completion. In *Proc. of SIGKDD*. 56, 67, 74

A. Vaswani, N. Shazeer, N. Parmar, L. Jones, J. Uszkoreit, A. N. Gomez, and L. Kaiser. 2017. Attention is all you need. In *Proc. of NIPS*, pages 5998–6008. 36, 39, 59, 60, 61, 79

P. Velickovic, G. Cucurull, A. Casanova, A. Romero, P. Lio, and Y. Bengio. 2018. Graph attention networks. In *Proc. of ICLR*. 39, 40, 59, 60, 78

P. Veličković, W. Fedus, W. L. Hamilton, P. Liò, Y. Bengio, and R. D. Hjelm. 2019. Deep graph infomax. In *Proc. of ICLR*. 56

O. Vinyals, M. Fortunato, and N. Jaitly. 2015. Pointer networks. In *Proc. of NIPS*, pages 2692–2700. 84

N. Wale, I. A. Watson, and G. Karypis. 2008. Comparison of descriptor spaces for chemical compound retrieval and classification. *Knowledge and Information Systems*, 14(3):347–375. DOI: 10.1007/s10115-007-0103-5 87

D. Wang, P. Cui, and W. Zhu. 2016. Structural deep network embedding. In *Proc. of SIGKDD*. DOI: 10.1145/2939672.2939753 56

P. Wang, J. Han, C. Li, and R. Pan. 2019a. Logic attention based neighborhood aggregation for inductive knowledge graph embedding. In *Proc. of AAAI*, 33:7152–7159. DOI: 10.1609/aaai.v33i01.33017152 72, 89

T. Wang, R. Liao, J. Ba, and S. Fidler. 2018a. NerveNet: Learning structured policy with graph neural networks. In *Proc. of ICLR*. 63

X. Wang, R. Girshick, A. Gupta, and K. He. 2018b. Non-local neural networks. In *Proc. of CVPR*, pages 7794–7803. DOI: 10.1109/cvpr.2018.00813 3, 59, 60, 61, 62, 64

X. Wang, Y. Ye, and A. Gupta. 2018c. Zero-shot recognition via semantic embeddings and knowledge graphs. In *Proc. of CVPR*, pages 6857–6866. DOI: 10.1109/cvpr.2018.00717 75, 76

X. Wang, H. Ji, C. Shi, B. Wang, Y. Ye, P. Cui, and P. S. Yu. 2019b. Heterogeneous graph attention network. In *Proc. of WWW*. DOI: 10.1145/3308558.3313562 48

Y. Wang, Y. Sun, Z. Liu, S. E. Sarma, M. M. Bronstein, and J. M. Solomon. 2018d. Dynamic graph CNN for learning on point clouds. *ArXiv Preprint ArXiv:1801.07829*. DOI: 10.1145/3326362 78

Z. Wang, T. Chen, J. S. J. Ren, W. Yu, H. Cheng, and L. Lin. 2018e. Deep reasoning with knowledge graph for social relationship understanding. In *Proc. of IJCAI*, pages 1021–1028. DOI: 10.24963/ijcai.2018/142 77

Z. Wang, Q. Lv, X. Lan, and Y. Zhang. 2018f. Cross-lingual knowledge graph alignment via graph convolutional networks. In *Proc. of EMNLP*, pages 349–357. DOI: 10.18653/v1/d18-1032 72

N. Watters, D. Zoran, T. Weber, P. Battaglia, R. Pascanu, and A. Tacchetti. 2017. Visual interaction networks: Learning a physics simulator from video. In *Proc. of NIPS*, pages 4539–4547. 59, 67

L. Wu, P. Sun, Y. Fu, R. Hong, X. Wang, and M. Wang. 2019a. A neural influence diffusion model for social recommendation. In *Proc. of SIGIR*. DOI: 10.1145/3331184.3331214 74

Q. Wu, H. Zhang, X. Gao, P. He, P. Weng, H. Gao, and G. Chen. 2019b. Dual graph attention networks for deep latent representation of multifaceted social effects in recommender systems. In *Proc. of WWW*, pages 2091–2102. ACM. DOI: 10.1145/3308558.3313442 74

Z. Wu, S. Pan, F. Chen, G. Long, C. Zhang, and P. S. Yu. 2019c. A comprehensive survey on graph neural networks. *ArXiv Preprint ArXiv:1901.00596*. 3

Z. Wu, S. Pan, G. Long, J. Jiang, and C. Zhang. 2019d. Graph waveNet for deep spatial-temporal graph modeling. *ArXiv Preprint ArXiv:1906.00121*. DOI: 10.24963/ijcai.2019/264 51

K. Xu, C. Li, Y. Tian, T. Sonobe, K. Kawarabayashi, and S. Jegelka. 2018. Representation learning on graphs with jumping knowledge networks. In *Proc. of ICML*, pages 5449–5458. 43, 44

K. Xu, L. Wang, M. Yu, Y. Feng, Y. Song, Z. Wang, and D. Yu. 2019a. Cross-lingual knowledge graph alignment via graph matching neural network. In *Proc. of ACL*. DOI: 10.18653/v1/p19-1304 72

N. Xu, P. Wang, L. Chen, J. Tao, and J. Zhao. 2019b. Mr-GNN: Multi-resolution and dual graph neural network for predicting structured entity interactions. In *Proc. of IJCAI*. DOI: 10.24963/ijcai.2019/551 70

S. Yan, Y. Xiong, and D. Lin. 2018. Spatial temporal graph convolutional networks for skeleton-based action recognition. In *Proc. of AAAI*. DOI: 10.1186/s13640-019-0476-x 51

B. Yang, W.-t. Yih, X. He, J. Gao, and L. Deng. 2015a. Embedding entities and relations for learning and inference in knowledge bases. In *Proc. of ICLR*. 71

C. Yang, Z. Liu, D. Zhao, M. Sun, and E. Y. Chang. 2015b. Network representation learning with rich text information. In *Proc. of IJCAI*, pages 2111–2117. 2

Z. Yang, W. W. Cohen, and R. Salakhutdinov. 2016. Revisiting semi-supervised learning with graph embeddings. *ArXiv Preprint ArXiv:1603.08861*. 87

L. Yao, C. Mao, and Y. Luo. 2019. Graph convolutional networks for text classification. In *Proc. of AAAI*, 33:7370–7377. DOI: 10.1609/aaai.v33i01.33017370 78

R. Ying, R. He, K. Chen, P. Eksombatchai, W. L. Hamilton, and J. Leskovec. 2018a. Graph convolutional neural networks for web-scale recommender systems. In *Proc. of SIGKDD*. DOI: 10.1145/3219819.3219890 53, 67, 74, 92

Z. Ying, J. You, C. Morris, X. Ren, W. Hamilton, and J. Leskovec. 2018b. Hierarchical graph representation learning with differentiable pooling. In *Proc. of NeurIPS*, pages 4805–4815. 55, 67

J. You, B. Liu, Z. Ying, V. Pande, and J. Leskovec. 2018a. Graph convolutional policy network for goal-directed molecular graph generation. In *Proc. of NeurIPS*, pages 6410–6421. 83

J. You, R. Ying, X. Ren, W. Hamilton, and J. Leskovec. 2018b. GraphRNN: Generating realistic graphs with deep auto-regressive models. In *Proc. of ICML*, pages 5694–5703. 83

B. Yu, H. Yin, and Z. Zhu. 2018a. Spatio-temporal graph convolutional networks: A deep learning framework for traffic forecasting. In *Proc. of IJCAI*. DOI: 10.24963/ijcai.2018/505 50

F. Yu and V. Koltun. 2015. Multi-scale context aggregation by dilated convolutions. *ArXiv Preprint ArXiv:1511.07122*. 45

W. Yu, C. Zheng, W. Cheng, C. C. Aggarwal, D. Song, B. Zong, H. Chen, and W. Wang. 2018b. Learning deep network representations with adversarially regularized autoencoders. In *Proc. of SIGKDD*. DOI: 10.1145/3219819.3220000 56

R. Zafarani and H. Liu, 2009. Social computing data repository at ASU. http://socialcomp uting.asu.edu 87

M. Zaheer, S. Kottur, S. Ravanbakhsh, B. Poczos, R. R. Salakhutdinov, and A. J. Smola. 2017. Deep sets. In *Proc. of NIPS*, pages 3391–3401. 59, 64

V. Zayats and M. Ostendorf. 2018. Conversation modeling on reddit using a graph-structured LSTM. *TACL*, 6:121–132. DOI: 10.1162/tacl_a_00009 35

D. Zhang, J. Yin, X. Zhu, and C. Zhang. 2018a. Network representation learning: A survey. *IEEE Transactions on Big Data*. DOI: 10.1109/tbdata.2018.2850013 2

F. Zhang, X. Liu, J. Tang, Y. Dong, P. Yao, J. Zhang, X. Gu, Y. Wang, B. Shao, R. Li, et al. 2019. OAG: Toward linking large-scale heterogeneous entity graphs. In *Proc. of SIGKDD*. DOI: 10.1145/3292500.3330785 72

J. Zhang, X. Shi, J. Xie, H. Ma, I. King, and D.-Y. Yeung. 2018b. GaAN: Gated attention networks for learning on large and spatiotemporal graphs. In *Proc. of UAI*. 40

Y. Zhang, Q. Liu, and L. Song. 2018c. Sentence-state LSTM for text representation. In *Proc. of ACL*, 1:317–327. DOI: 10.18653/v1/p18-1030 36, 75, 78, 79

Y. Zhang, P. Qi, and C. D. Manning. 2018d. Graph convolution over pruned dependency trees improves relation extraction. In *Proc. of EMNLP*, pages 2205–2215. DOI: 10.18653/v1/d18-1244 79

Y. Zhang, Y. Xiong, X. Kong, S. Li, J. Mi, and Y. Zhu. 2018e. Deep collective classification in heterogeneous information networks. In *Proc. of WWW*, pages 399–408. DOI: 10.1145/3178876.3186106 48

Z. Zhang, P. Cui, and W. Zhu. 2018f. Deep learning on graphs: A survey. *ArXiv Preprint ArXiv:1812.04202*. 3

J. Zhou, X. Han, C. Yang, Z. Liu, L. Wang, C. Li, and M. Sun. 2019. Gear: Graph-based evidence aggregating and reasoning for fact verification. In *Proc. of ACL*. DOI: 10.18653/v1/p19-1085 81, 82

H. Zhu, Y. Lin, Z. Liu, J. Fu, T.-S. Chua, and M. Sun. 2019a. Graph neural networks with generated parameters for relation extraction. In *Proc. of ACL*. DOI: 10.18653/v1/p19-1128 79

R. Zhu, K. Zhao, H. Yang, W. Lin, C. Zhou, B. Ai, Y. Li, and J. Zhou. 2019b. Aligraph: A comprehensive graph neural network platform. *arXiv preprint arXiv:1902.087.30*. 89

Z. Zhu, S. Xu, M. Qu, and J. Tang. 2019c. Graphite: A high-performance cpu-gpu hybrid system for node embedding. In *The World Wide Web Conference*, pages 2494–2504, ACM. 89

C. Zhuang and Q. Ma. 2018. Dual graph convolutional networks for graph-based semi-supervised classification. In *Proc. of WWW*. DOI: 10.1145/3178876.3186116 28

J. G. Zilly, R. K. Srivastava, J. Koutnik, and J. Schmidhuber. 2016. Recurrent highway networks. In *Proc. of ICML*, pages 4189–4198. 43

M. Zitnik and J. Leskovec. 2017. Predicting multicellular function through multi-layer tissue networks. *Bioinformatics*, 33(14):i190–i198. DOI: 10.1093/bioinformatics/btx252 87

M. Zitnik, M. Agrawal, and J. Leskovec. 2018. Modeling polypharmacy side effects with graph convolutional networks. *Bioinformatics*, 34(13):i457–i466. DOI: 10.1093/bioinformatics/bty294 70

Authors' Biographies

ZHIYUAN LIU

Zhiyuan Liu is an associate professor in the Department of Computer Science and Technology, Tsinghua University. He got his B.E. in 2006 and his Ph.D. in 2011 from the Department of Computer Science and Technology, Tsinghua University. His research interests are natural language processing and social computation. He has published over 60 papers in international journals and conferences, including IJCAI, AAAI, ACL, and EMNLP.

JIE ZHOU

Jie Zhou is a second-year Master's student of the Department of Computer Science and Technology, Tsinghua University. He got his B.E. from Tsinghua University in 2016. His research interests include graph neural networks and natural language processing.

Printed in the United States
by Baker & Taylor Publisher Services